NIGEL HENBEST & HEATHER COUPER

EXTREME UNIVERSE

NIGEL HENBEST & HEATHER COUPER

EXTREME UNIVERSE

First published 2001 by Channel 4 Books
an imprint of ≈ Pan Macmillan Ltd
Pan Macmillan, 20 New Wharf Road, London, N1 9RR
Basingstoke and Oxford
Associated companies throughout the world
www.panmacmillan.com

ISBN 07522 6163 0

A CIP catalogue record for this book is available from
the British Library.

Design by DW Design, London

Printed and bound in Italy by New Interlitho

This book accompanies the television series *Extreme Universe* made
by Pioneer Productions for Channel 4.
Executive Producer: Stuart Carter
Series Producer: Martin Mortimore
Scriptwriter: Nigel Henbest

contents

The current

Worlds

chapter one

total of extrasolar planets stands at sixty-three

Beyond

Didier Queloz was a worried man. 'I must admit that when I first saw data that was not in agreement with the past data on the star,' he recalls, 'I was very puzzled – and I really thought, "Oh my God, something is very wrong with the instrument."' The twenty-nine-year-old graduate student could hardly have foreseen at the time that his 'faulty instrument' was telling the truth – and that he was about to let loose a revolution in cutting-edge science. Although he didn't know it then, Didier Queloz had become the first person in history to discover a planet orbiting another star.

Searching for planets around other stars has become a Holy Grail for astronomers. Until the closing years of the last millennium, we knew only of the nine planets orbiting our Sun – the worlds of our own Solar System. And we knew that on at least one of those worlds there was abundant life. Looking for planets beyond the Solar System was driven by two motives: to probe the cosmos, and – in a sense – to look for ourselves. At the back of every planet hunter's mind is the thought: 'Can we really be alone in this enormous Universe?'

ABOVE *Didier Queloz at the Observatoire de Haute Provence, where he discovered the planet orbiting the star 51 Pegasi.*

OPPOSITE *The rival planet-hunting team from the Lick Observatory: Paul Butler, Geoff Marcy and Debra Fischer.*

But it's not an easy task. 'Right now, it's impossible even with our most powerful telescopes to see a planet,' explains Debra Fischer from the Lick Observatory, part of the University of California. 'The planet reflects such a tiny amount of starlight. It's like a little firefly buzzing around in the headlights of an oncoming car.' So astronomers need to think laterally. 'It would be great if we could go up and take a picture of a star and see the planet sitting right next to it, but instead we have to be a little more clever,' she continues. 'We know that, as the planet orbits a star, the star has the gravitational leash that holds the planet in its orbit. But as the planet goes round, it also tugs on the star.' Her colleague Geoff Marcy adds, 'We watch the star to see if it wobbles in space. We're simply watching to see if stars are stationary or wobbling.'

The problem lies in the extent of the wobbling: even a massive planet like Jupiter can hardly perturb the mighty Sun. But challenges like this have never stopped people trying. One stalwart of the wobble technique was Peter van de Kamp, who worked at the Sproul Observatory in Allegheny, Pennsylvania. For fifty years, he painstakingly catalogued the wobbles he thought he had detected in a nearby red dwarf called Barnard's Star. In the days before computers, electronics and high-precision instrumentation, it was a mighty effort. It involved taking pictures of the star on old-fashioned photographic plates, year in, year out – and then trying to measure displacements as small as one-thousandth of a millimetre.

The wobbles he measured convinced him that Barnard's Star had two planets in tow, with masses similar to Jupiter and Saturn in our own Solar System. But after van de Kamp retired, his successors pored over the material he had gathered. They also compared it with measurements of the performance of the telescope. The conclusion was devastating: the wobbles in Barnard's Star exactly matched the oscillations in the telescope. Van de Kamp was using a long, old-fashioned lens telescope which flexed as it pointed to different areas of the sky. So, while the precision of his data was second to none, it transpired that he was measuring the wrong thing. Instead of detecting the pull of planets around Barnard's Star, he was carefully logging the pull of gravity on an ancient telescope.

Nonetheless, a new generation of planet hunters was inspired by van de Kamp's work. They knew that the key to discovery was precision – and by the late 1980s, technology was beginning to come on side. Paul Butler, who works with Geoff Marcy and Debra Fischer, describes what they were up against. 'The Sun is a million miles in diameter, so the typical stars that we're looking at are of order a million miles in diameter. And they're at these unbelievable distances – a hundred light years would be typical.' A hundred light years is nearly 600 million million miles. 'And so we're looking at these stars a hundred light years away, a million miles across, and we need to be able to tell if they're moving towards us or away from us at a precision of about three metres a second – which is sort of a fast walk or a very slow trot. When you think about it, it's really kind of mind-numbing.

'In particular, a star is a boiling ball of gas – it's not a hard surface. When we started this project, we didn't know how intrinsically stable stars were. It might be that stars have jitters on the ten, twenty or fifty feet per second level – it simply wasn't known.'

The magic bullet in a planet finder's armoury is the technique of spectroscopy. It works like this: just as crystal glasses break up sunlight into all its colours, so a spectrograph can spread out starlight into a rainbow. What you're seeing are the 'notes' of which that light is composed – red for the low notes (or long wavelengths), blue for the high notes (short wavelengths).

'The way we detect the wobble of a star is really simple in principle, difficult in practice,' admits Geoff Marcy. 'A star emits all the colours of the rainbow – blue, green, yellow and red – and when a star is moving at you, those colours of the rainbow shift towards slightly adjacent colours. And when the star is moving away from you, the colours of the rainbow shift towards the other end of this spectrum.

'At the back of the telescope, we put a spectrograph where your eyeball would normally go, and see if the colours are shifting back and forth – that's the Doppler Effect.' Adds Debra Fischer: 'The Doppler Effect is something that people are more familiar with in sound. You hear the pitch of the sound change as a car or train goes past you, because the wavelength changes.'

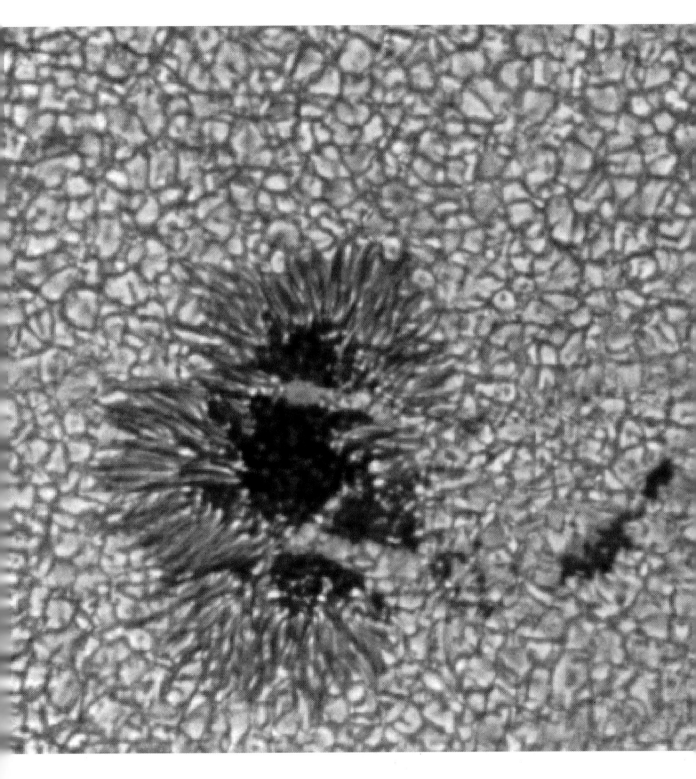

ABOVE *A huge sunspot – bigger than Earth – dominates this close-up of our local star. All around, the gaseous surface churns and boils. Astronomers monitoring stars for planets must convince themselves they are detecting real 'wobbles' rather than the star's surface churning.*

'Spectroscopy is a wonderful, wonderful science,' enthuses Paul Butler. 'From this little teeny beam of light, not only can we get the speed of the star in space, but also its temperature, the surface pressure, and the chemical composition of the star. It's really astonishing the amount of information that rides on those little beams of light.'

But Geoff Marcy's warning that making the measurements would be difficult in practice proved to be all too true. 'We spent eight years basically in hell,' recalls Paul Butler with a shudder. 'We couldn't figure out why our measurement errors were typically fifty or a hundred metres a second instead of under ten metres a second. And the problem turned out to be very, very small changes in the spectrograph, caused by the temperature and pressure going up and down – because we live in the real world. It took us eight years to write a piece of software that allowed us to calibrate that out and get down to the fundamental limit.'

Meanwhile, on the other side of the Atlantic, another team of planet hunters was swinging into action. Michel Mayor of the Geneva Observatory, and his graduate student Didier Queloz, were having the same teething problems as the Lick Observatory team. 'It was early '90,' Queloz reminisces. 'That's when I started my thesis. After hard work, a lot of software, and a lot of debugging, we had the instrument on the telescope.'

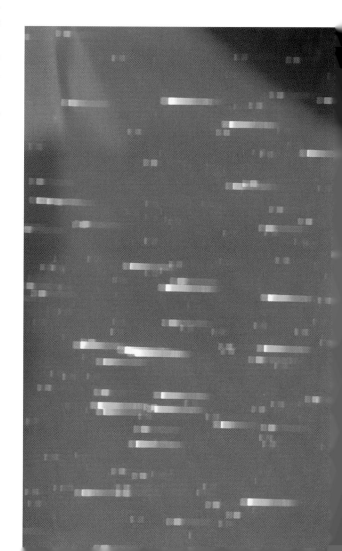

In late 1994, satisfied that their spectrograph, 'Elodie', could measure wobbles as small as ten metres per second, the team started a systematic search for planets around 130 nearby Sun-like stars. 'No one expected that we would find a planet from the very beginning,' explains Queloz, 'and to give you an example, Michel Mayor decided to have a six-month sabbatical on Hawaii four months after the beginning of the programme.'

In the early spring of 1995, Queloz – alone – continued his search with Elodie at the Observatoire de Haute Provence. This cluster of telescopes, perched high in the wooded Provençal countryside close to the wine-growing region of the Côtes du Luberon, is not one of the world's prime observing sites. And the telescope Queloz was using – with a mirror six feet across – is

hardly one of the world's largest, as he himself admits. 'In the astrophysical business, telescopes mean big guys – six metre, eight metre, ten metre. But it's very difficult to get, let's say, two or three months of telescope time per year on a big telescope, because there are a lot of people who want to use it. If you use a smaller telescope, you can get one week a month. So you have a kind of dedicated facility.'

One of the stars Queloz monitored was 51 Pegasi, fifty light years away and visible in the night sky in the constellation of Pegasus, the flying horse. And that's when his problems started. 'Every night I was observing that star, I got a different measurement for its speed. Then I said, "Oh my God, that's not the same value" – it was very, very worrisome for me at first hand.'

Queloz – in common with all his fellow planet hunters – was expecting a very slow change in the speed of the star wobbles. They were all aware that even with their new precision techniques they would be unable to detect planets less massive than Jupiter and Saturn. And the layout of the Solar System – not to mention the theories of planetary formation – has giant planets like these placed at some considerable distance from their parent star. This far out, it takes a long time for a world to trek around its star. Jupiter takes 11.9 years to orbit the Sun once; Saturn almost thirty. These were the sorts of periods Queloz was looking for. But he was picking up a variation of just four *days*.

'I saw something having this variation of four days. Well, I had seen no other explanation other than there was something orbiting it. Then I said to myself, "Keep focused," and I had what I said was a kind of naive way of thinking, like: "OK – Jupiter is eleven years, but why not? Maybe there may be a planet on this." If I had been older

ABOVE *An old constellation print of Pegasus, the winged horse. The first star to have a planet discovered, 51 Pegasi – visible to the unaided eye – is highlighted at the centre.*

OPPOSITE *To detect planets, astronomers spread out a star's light into a spectrum – a miniature rainbow. A slight shift in the colours towards or away from the ends of the spectrum reveals if it is wobbling under the influence of orbiting planets.*

with a lot of experience in the planet field, I would have said, "Oh no, it cannot be a planet, there is no chance – so maybe I should not focus on 51 Peg." '

But focus on 51 Peg he did. With the data he had acquired, he computed an orbit for a planet that orbited its star in just 4.2 days (even Mercury, the innermost planet in the Solar System, takes eighty-eight days to circle the Sun). At the end of the observing season, late in March 1995, he made the critical observations which proved that his 'naive' instincts seemed to be right: 51 Peg was apparently being circled by a planet half the mass of Jupiter that whizzed round its star so close in that it was almost touching the star's surface.

Queloz had never encountered a scenario so bizarre. And it didn't help that his supervisor Michel Mayor – whom he urgently wanted to run things past – was somewhere on the other side of planet Earth. 'I sent him an e-mail saying, "Look, Michel, I think we've found something interesting." He never asked me, "Are you sure about the data?" And he never did doubt my work. He said, "I believe what you've done," and he really trusted me a lot.'

Meanwhile, Queloz – with his controversial discovery yet to be confirmed – was all on his own until his supervisor returned from Hawaii. Being the only person in the world harbouring such an explosive secret was hard. 'I remember the feeling I had, I remember being very afraid. At the end of every night I said, "OK, maybe I've made a big mistake and I have to sleep – I am very tired."

'When I woke up in the afternoon, I said, "Let's see what's up tonight." And then, suddenly, clouds. I said, "Oh no, I cannot observe tonight, so I will be unable to confirm what I did." And this is part of the whole process – a lot of emotion. And the most interesting point of that is that I was single: I couldn't share it. Michel was not there; he was very far away. It was just me and the telescope and the data. And I had this all to myself for quite a while.'

At the end of March, 51 Peg disappeared into daylight skies, not to reappear until the end of the summer. So there was little that Queloz could do but wait. Fortunately Michel Mayor returned to Geneva in April, and was able to give his student much-needed support. Queloz recalls his supervisor's wise counsel. 'He said, "It may be a planet, but let's be cool. Let's wait until July to make sure what we have." '

Both astronomers returned to Provence in July when 51 Peg re-emerged in dark skies. To Queloz's delight, the new observations confirmed what he had found in the spring – and he was ecstatic to share the discovery with his supervisor. 'Michel got the thrill when he came here in July with me, and we got confirmation of the data, and then we were completely excited. Scary, as well. I mean, we were excited and scary. Excited because we knew we had something great, but scary because we knew we had a planet that was not the planet that everybody was expecting. That was a little scary – maybe I should say a lot scary.'

Queloz and Mayor spent three months working on the data to make sure that they really had discovered a planet – ruling out other possible interpretations such as the rotation of the star itself. But on 6 October 1995 they felt confident enough to travel to France and announce their result. 'I would say that most of the people in the room didn't believe us, I'm pretty sure,' says Queloz. 'They said, "Well, that's crazy." But some people thought, "OK, maybe – but we'll see. We have to see if somebody else could confirm it."'

The news caused waves right across the world. Over in California, Doug Lin, theoretician at the Lick Observatory remembers his reaction. 'I thought, "This sounds very interesting – but is the discovery real?" The general reaction at the time was that there's got to be a mistake in the data – this couldn't be due to a planet, because it looks so different from our own Solar System. And there had been lots of false claims in the past. So I picked up the phone and called my friend Geoff Marcy. He'd just come down from the mountain ten minutes earlier.'

Not too coincidentally – having also heard the announcement from France – Marcy and his colleague Paul Butler had also been looking at 51 Peg, using the ten-foot-diameter telescope down the road atop Mount Hamilton. 'I asked him about the discovery of this 51 Pegasus,' recalls Lin, 'and I asked him if there was indeed a planet. And he said – with a great deal of excitement – yes indeed! He said that their data absolutely confirmed the Swiss discovery.'

'So I asked, "Could you tell me more?" He said, "No, I have to go into a press conference." So I asked him to give me just two numbers, the period of the orbit and the mass. In the next hour, I very busily worked out the numbers and came up with a theory of how these planets could form at a relatively large distance, migrate in, and stop so close to the host star.'

ABOVE *Newspaper headlines around the world sensationalized the discovery of the planet around 51 Pegasi.*

Marcy, who had been on the trail of extrasolar planets for eight years by then, was staggered about the discovery. 'We were so shocked that we didn't believe it was a planet at first,' he admitted. 'We had to go to the telescope ourselves to confirm it. And the reason it was a shock was that this planet goes around the star 51 Peg in only four days.'

'We were sceptical,' agrees his colleague Paul Butler. 'We thought – "four days! This must be crazy – we've never heard of such a thing", and I believe there was a lot of scepticism in the community at large. The one thing that it had seriously going for it was the reputation of Michel Mayor. He's a brilliant astrophysicist, and he has a long and very proud record of just doing wonderful science.

'And so that was the one thing that really made us take this thing very seriously. Fortunately, in our case, we had telescope time about ten days later, and we had four nights. And this was a four-day period orbit.'

But technology was still not quite on their side. 'In the dark ages of 1995, computers were a lot slower than they are now,' recalls Butler. 'When we started the data analysis, it took all day to do just one star. After the end of the observing run, we got back – it's about three in the afternoon – we haven't slept, we're dead tired, so we agreed to meet in our office that night at about ten o'clock. And so I went home and crashed, and I think Geoff did the same.

'We met back in the office that night and the data was still coming in. When we saw that fourth night come in, it did exactly as predicted. It was stunning – we both felt shivers. In addition to Mayor and Queloz, Geoff and I knew this was right, and nobody else on the planet knew at this point.

'The first issue is, what should we do? Since we were able to confirm the result, and we knew that they were right, we thought we should just say so. There was so much scepticism at the time about this claim that we thought we should just make it publicly available. So we took our plot of the four nights of data, and simply put it on the Internet – we just made it available to the whole world.'

At that point, Geoff Marcy and Paul Butler came to a startling realization. They had been gathering data from stars for many years, assuming that any planets out there would take many years to orbit their star – not days. So they re-analysed their accumulated data, and, within two weeks of the announcement of the finding of the planet around 51 Peg, they were able to report two new planets – again with orbital periods much shorter than expected – that had hitherto gone undetected.

Geoff Marcy relives the moment. 'The high point for us was when we detected our first two planets. Within about two weeks we found the planets around 70 Virginis and 47 Ursae Majoris, and you have to imagine, here we'd been working for ten years solid, every day, seven days a week, looking for the wobble of a star. And finally, one day, we looked at our data and there was the wobble. To see that on the computer screen was by far the most momentous moment in my entire life.'

Now the field was wide open. 'I mean, it's a shock to everyone that you don't need eleven years to detect a planet,' observes Didier Queloz. 'You really need one week. So Geoff Marcy's team, who started four or five years before us, realized that maybe in their data they had planets. And it was a big push for many teams to go into this business, because it was much easier than they had thought – at least on the timescale, although not on the accuracy. You need to be accurate – you need to find the trick to reach that.'

Geoff Marcy recalls the heady, early days of the mid-nineties planet searches. 'There was a bit of a race going on, and I'm almost embarrassed to admit it – there were about half a dozen different research teams hunting for planets then. It was like a gold rush in the sense that we were realizing for the first time that there were planets that took only a few weeks or months to go around their star, and it's like a gold vein in a mine. Now we knew where to look for planets, and that some of them might be in fairly close. And we looked and – sure enough – we found dozens of them.'

ABOVE *Our Milky Way Galaxy alone contains around 200,000 million stars. With the latest discoveries of new worlds, astronomers are becoming confident that planets are common in the Universe.*

The current total of extrasolar planets stands at sixty-three ('We have detected thirty-four of them,' says Didier Queloz, proudly). Marcy and Butler's group is very slightly behind, but they have many more suspected planets in the can. And the fact that they're based at the Lick Observatory – owned by the University of California – buys them time on the world's biggest telescope, the Keck, also run by the University.

The Keck is a modern-day Cyclops – and now there are two of them. These twin telescopes are perched on the barren, moonlike summit of Mauna Kea on the Big Island of Hawaii, nearly fourteen thousand feet up. The Keck Telescopes' 'eyes' are gigantic mirrors thirty-three feet across – offering a light-collecting area equivalent to half a tennis court. From this clear, dry and disconcertingly airless environment – where every exertion leaves you tired and breathless – the Kecks have one of the greatest views of the Universe on Earth.

To Geoff Marcy, using the Keck is an enormous privilege. 'When we're using the Keck Telescope, it's profoundly moving for us. I almost feel tears coming to my eyes – it's so lucky that we humans have this marvellous machine to search for planets.

'What happens most commonly is that we are pointing to a new star, and that we're going to observe it for ten minutes. One of us is typing away on the computer, bringing up the latest data we have on the star – does it have a planet or not? And we sometimes realize that there's a hint of a planet in the data, we can see the star wobbling a little bit. So we very excitedly observe that star, and often go back an hour later. Then we get into a sort of playful argument about whether that star has planets or not, based on the data that we have. By arguing about it, we suss out the pros and cons about the existence of the planet – and we learn by talking to each other much more than we could learn by just thinking ourselves.'

'It's a really nice feeling, observing,' adds Paul Butler. 'I think about three or four in the morning, you have the sense that it's you and the sky. The sky is yours and everyone's asleep, the whole world's asleep, and you've got the world's biggest telescope at your control. It's a lot of fun.'

The trick of being a successful planet hunter is to be able to select the right stars to observe in the first place. Geoff Marcy admits that he took a bit of time getting this right. At first, he pointed the telescope at all kinds of different stars, without success. Then his team twigged that stars like the Sun were more likely to come up with the goods – but, as it turned out, they hadn't got it entirely right.

Then – out of the blue – they acquired a secret weapon: an undergraduate student at Sussex University. 'I met Kevin Apps in the strangest way,' muses Geoff Marcy. 'One day here at the Keck Telescope I was reading my e-mail, and here comes one from another fan – that's what it seemed like. He said, "I would like to be sent your target list of stars." And I thought, "The audacity of this young kid, asking me for our precious list of target stars that we've culled out of the billions of stars in our

OPPOSITE *The twin domes of the giant Keck Telescopes on Hawaii. It's here – at fourteen thousand feet – that Marcy's team conducts its search for planets around other stars.*

Galaxy." And he went on in his e-mail to say he would like to assess their characteristics, and I thought – "Well, that sounds very thoughtful."

'And to my surprise, two days later, he came back with an e-mail saying, "I'm sorry to report to you that some of your stars are unacceptable." I thought, "How can a mere amateur be telling me which stars we should be observing?"

'Turned out he was absolutely right. Some of the stars we were observing were binary stars, two stars going round each other, so there could never be a planet in there. Some of them had rapid spin rates that would be terrible for us as well. We adopted his suggestions one hundred per cent, and indeed nowadays he's the one who supplies us with the new stars which replace the old ones we throw out.'

Kevin Apps was ahead of the pack. An enthusiastic amateur astronomer, reading astrophysics at university, he had come across precise new measurements on stars acquired by a revolutionary new satellite called Hipparcos. He was also blessed with a photographic recall of data – having the ability to remember arcane facts like the spin-rate of a star, what it's made of, and how far away it is.

Today, young Kevin Apps is a treasured member of Marcy's team – even though he is still a student based in England. 'Kevin has saved us thousands of hours of telescope time,' admits Marcy. 'We would have been wasting time – at a thousand dollars a minute, or whatever it is for the Keck Telescope – looking at stars that had no chance for us to find a planet.'

In 1996, Marcy's team found a planet circling the star upsilon Andromedae. Like most of the stars in the team's sample, it's visible to the unaided eye: 'You can go out at night and look at it,' says Marcy. 'Now, the discovery of the planet was all very nice, but we continued to watch the star. To our surprise – and frankly, we were a little worried – the star didn't obey the predicted wobble.

'At first, we thought we'd made a terrible mistake. But then we started to realize that the star was wobbling in addition to its original wobble, a second wobble – and then as the years went by, a third wobble. The star was wobbling in a kind of curlicue pattern due to the fact that there were three planets going around it. They were yanking it in different directions, like a dog owner with three dogs on three different leashes.'

This time, the team had not just discovered a planet – they had discovered the first planetary system in our Galaxy outside the Solar System. And they didn't quite believe it. Debra Fischer, who, unbeknown to Geoff Marcy, did an independent analysis of the data, remembers her reaction. 'When I realized that there were three planets in the system and not just two, it was astonishing. My first fear was that, you know, I was wrong and that I would look foolish when this news came out.'

Meanwhile, Geoff Marcy was experiencing similar emotions. 'It was a shocking moment. I remember one evening I was at the computer and I couldn't get the data to make sense. And in the wee hours of the morning I finally hit on it, and realized I needed to have three planets to understand this star, not two. I remember next day talking to my colleagues about it, and I didn't want to tell them the answer. I said, "Go try this on your own and see what you find," 'cause it was too fantastic to believe.'

When they finally suspended their disbelief, the team was euphoric. 'It was profoundly moving,' recalls Debra Fischer. 'If you could have a star that made three of these giant Jupiter-like planets, then the whole process of making planets had to be easy – and by implication, all stars are born with an array of planets around them.'

But the dozens of planets that have now been discovered around other stars all have one thing in common: no one has actually seen any of them. Their presence has been inferred purely from the gravitational tug they exert on the parent star. But for astronomers, this isn't a problem. 'Every once in a while in astronomy, a new phenomenon is discovered which you can't actually see,' observes Andrew Collier-Cameron, a planet hunter at the University of St Andrews in Scotland.

'A very good example of this is black holes. Nobody knew what a black hole ought to look like. So the first step is to formulate a theory, and to use your knowledge of physics to predict what we might detect – like, how fast material would be orbiting as it fell into a black hole, how much energy it would lose, and how hot it would be. It's the same with planets around other stars.'

Canadian graduate student Dave Charbonneau understands how important it is to actually see an extrasolar planet. 'I think what we'd all like to see – astronomers and the general public alike – would be to see a picture of a planet orbiting another star. Unfortunately, we can't do that. The planets are overwhelmed by these massive brilliant stars right by them.'

But twenty-seven-year-old Charbonneau, based at the Harvard–Smithsonian Center for Astrophysics, has come the closest ever to seeing a planet around another star – or, rather, seeing an 'anti-planet'. He was able to infer the existence of a new world by the absence of light from its parent star.

'I was working with my advisor Bob Noyes,' recounts Charbonneau, 'and we were trying to come up with a plan for a good thesis project. We knew at the time that

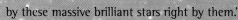

ABOVE *Dave Charbonneau was a twenty-five-year-old student when he discovered a planet using a home-made telescope. Now he has been awarded observing time on the Hubble Space Telescope.*

WORLDS BEYOND

21

there had been dozens of planets detected by the wobble technique, and we wanted to come up with an idea that would allow us to learn something about these planets that the wobble technique just couldn't tell you.

'So we racked our brains and realized that the transit technique would be the most exciting one to follow. The basic idea is that when a planet orbits a star – if you're looking at the right planet and the right star with the right tilt – the planet passes in front of the star as it orbits. Then what you'll see is a little dimming of the light from the star for about three to four hours. So you can measure the size of the planet, and its mass – and those two things together would tell you the density, and what the planet's made of.

'It was something that we really thought was worthwhile, and so what Bob did was recommend that I came out to Colorado to work with Tim Brown, who had just completed the STARE Telescope.'

Tim Brown, of the National Center for Atmospheric Research in Boulder, Colorado, is a devoted telescope builder – making instruments both for science and for fun. A few years back, he came up with the idea of constructing a dedicated telescope to look for transits. Instead of the large telescopes involved in searching for wobbles, STARE – which stands for Stellar Astrophysics and Research on Exoplanets – has a mirror a mere four inches across.

'Prior to getting into a car to drive out to Boulder to work with Tim, I spoke to him on the phone to get up to speed on the operation of the STARE Telescope,' explains Charbonneau. 'We figured that the first thing to do would be to look at a number of stars that were known to have planets, and to measure the brightness of the stars to see – if we were fortunate – where the planet would go in front of the star and make a little eclipse.

'Dave Latham, in my university department – working with Michel Mayor and the Geneva group – had made a number of measurements of a star called HD 209458. They'd inferred that it had a planet orbiting in a very tight orbit, every three and a half days. There was a much better chance that the planet would go in front of the star and allow us to make a transit measurement. So in August 1999, I drove out west, met up with Tim, and we set about making observations of that one star.

'We didn't analyse the data for more than a month after we took it, and so when I was sitting at my computer finally looking at all this data we'd gathered in September, and I saw the first indication of a transit, it was very, very exciting. I think my first reaction was, this is too good to be true. So I went and checked the other night of data we had from a week later, to see if it showed the transit at exactly the right time. And then, when in those data, I saw the transit emerge once again, I had to sit back in my chair for a moment.'

Because Charbonneau and Brown knew the distance of the planet from its star, its mass, and how much light it absorbed when it crossed in front, they could work out what kind of beast it was. It was a world like Jupiter – but with a difference. 'Although it was similar to Jupiter in mass,' relates Charbonneau, 'it was significantly larger. It had been puffed up due to the proximity to the star and all the heat that was hitting it.'

The planet-hunting community collectively gave young Charbonneau an enormous pat on the back. 'Oh, the transit – that's a great discovery,' enthuses Didier Queloz. 'It gives you another way to detect a planet – another technique. That's what we call complete confirmation. And that's good – that's great.'

'The transit planet is very important,' adds Andrew Collier-Cameron, 'because up to the point where it was discovered, nobody knew whether these planets were genuine gas giants. There was a possibility that they could have been just lumps of rock, although it was hard to see how so much rock could form so close to the parent star. The transit immediately told us that this planet was about thirty per cent bigger than Jupiter, and there's no way that you can have a rock that size.'

Dave Charbonneau heaps his praises on the vigorous new discipline of planet hunting that afforded him the opportunity to make such a great advance. 'The whole field is only a few years old, and so it's possible for even graduate students to jump in and start doing exciting science.'

'I think it's an unparalleled time to be working in astronomy right now,' agrees Andrew Collier-Cameron. 'Quite apart from the advances that are being made in the study of the large-scale structure of the Universe, the last five years – this roller-coaster ride in the discovery of extrasolar planets – has been incredibly exciting to anybody like myself who, as a kid, grew up on a more or less steady diet of science fiction.'

'We're seeing a completely new breakthrough in science – like the discovery of quantum mechanics at around the turn of last century,' adds Doug Lin. 'It's like the revolution brought about by all those Internet companies. Only five or six years ago, when you talked about dotcoms, people on the street would say, "What's dotcom?" And today, life would be unthinkable without access to the Internet.'

'I think about why I've spent fifteen years of my life dedicated day and night, seven days a week to hunting for planets,' muses Geoff Marcy. 'Part of it is my ego, actually. And it's a little embarrassing to say, but I would love to make a contribution to science. I love science. I love human knowledge of any sort.

'I think of the glorious advances we humans have made in just two million years since we became a species. If it's at all possible, I'd like to make a tiny contribution to the grand history of human invention, and discovering planets would be what I would hope to offer.

'I'm a little worried that I'm too manic about it. Maybe I should just live my life and not worry about making a contribution, but I really want to do that. It'll make me feel like my life is complete, and it'll serve to give me a little bit of a sense of immortality. I'll feel that something I helped to discover is going to live on for hundreds and thousands of years as a precious discovery for all humans.'

Heavenly

quickly acquired a name for themselves – hot Jupiters

Bodies?

With the discovery of planets around other stars, most astronomers expected to feel content. A missing piece of the cosmic jigsaw puzzle had at last been found: planets appeared to be common, and our Solar System was no longer unique. Or – was it? From the moment Didier Queloz discovered the planet orbiting 51 Peg, the planet-hunting community realized that they had opened a can of worms.

'There's been one common theme that's run through the last five years,' observes Debra Fischer, 'that these extrasolar planets have been surprises to us in every way. The first one was a shock because, here it was – a half Jupiter-mass in a four-day orbit around its star. So for that planet, one year is four days.

'And there was a moment of stunned silence in the astronomical community as we said, "Wait, that can't be. That planet couldn't have formed there, how did it get there? If it did move into this position, what stopped it, why didn't it go all the way

ABOVE *Rogue worlds: many of the planets orbiting other stars are 'hot Jupiters' – vast worlds so close to their parent star that life would be impossible.*

OPPOSITE *Planets are born in dusty discs surrounding stars, like this ring around epsilon Eridani. But with the discovery of wacky new worlds, astronomers are realizing that they still have a lot more to learn about the formation process.*

into the star, isn't it so hot that the whole atmosphere of the planet would be blown away?" So that one discovery launched a hundred questions, and I think every new discovery has similarly generated a cascade of questions.'

The new discoveries meant that the ideas about how planets are born had been thrown into the melting pot. And this was at a time when astronomers were becoming increasingly confident that they'd got the right picture. Many stars, as they form, are surrounded by swirling discs of material which are construction sites for young worlds. Planets are a natural by-product of starbirth, and astronomers assumed that all planetary systems would turn out to look roughly similar.

'The discs around these stars contain all sorts of elements,' explains planetary scientist John Spencer of the Lowell Observatory in Arizona. 'These include iron and the makings of rock, and these will condense out as the nebula cools and form little dust grains. These then stick together to make larger objects, which will eventually start colliding with each other. They gradually build up into bigger and bigger things – asteroid-sized, then moon-sized, and eventually planet-sized.

'In the inner part of the Solar System, we think that's the main thing that happened. But in the outer part, it was colder, and ice could condense as well. And in the disc, there are a lot more of the constituents of ice than the constituents of rock. So you can make much bigger planets out there, and once a planet gets big enough, it starts sucking in gas from the disc by its own gravity. So the outer planets gather all this extra hydrogen and helium and become the gas giants. The inner planets never got big enough to do that.'

Years of computer-modelling of how planetary systems form came up with consistent results: small worlds orbiting close in to a star, giant worlds on the

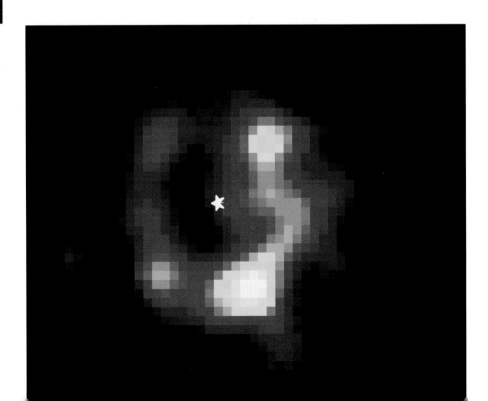

BELOW *Montage showing the worlds of our Solar System, to scale. Jupiter is big enough to contain all the other planets put together.*

BELOW CENTRE *Rocky Mercury is the closest planet to the Sun, and circles it in only eighty-eight days. It has been battered into submission by meteorite impacts.*

outskirts. In fact, the giant planets – especially Jupiter – are so huge that they dominate the whole scene. 'Isaac Asimov once said that the Solar System consisted of Jupiter plus debris,' recalls John Spencer, 'which is a rather extreme way of putting it. But Jupiter does have the bulk of the mass in the Solar System, and it really is the most important planet in terms of controlling what happens there.

'Looking at the Solar System from the outside, you would hardly notice the Earth – it's three hundred times less massive than Jupiter, and ten times smaller in radius. It's like a wimpy little pebble.'

What of the other wimpy pebbles? 'Mercury is the closest planet to the Sun, and it looks like the Moon,' Spencer continues. 'It's smaller than the other terrestrial planets, and, being so close to the Sun, it hasn't been able to hold on an atmosphere at all. Mercury and the Moon have been too small to sustain geological activity because they have no heat in their interiors – they're basically geologically dead, and all they do is sit there and get hit by asteroids and comets. And so Mercury is just a big cratered bowl of rock with no atmosphere, no water, and certainly no prospects of life.'

After Mercury comes Venus. 'The planet Venus is the closest to the Earth in size. But it's utterly different from our planet. Its atmosphere is about a hundred times thicker than Earth's, and it's mostly made of carbon dioxide. Because of the Greenhouse Effect trapping heat, it's also intensely hot – about nine hundred degrees Fahrenheit on the surface – and extremely dry. On Earth it was just cool enough for liquid water to form, and that allowed chemical reactions that sucked the carbon

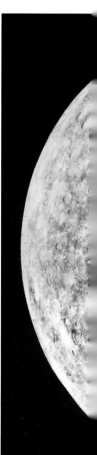

dioxide out of the atmosphere to make limestone and other rocks. They stored the carbon dioxide beneath the surface, and so the atmosphere never got that thick and never got that hot.

'Venus was just too close to the Sun for that to happen. It really gives you a feel for just how precarious the balance is that's required to sustain life. Venus just didn't quite make it.

'In the big picture of the Solar System, you wouldn't notice the Earth. Although it's the biggest piece of rock there, it's nothing like the size of Jupiter – it's a kind of afterthought. Of course for us, it's the most important place, and anyone looking for life would look for it here – we're close enough to the Sun for warmth, and we have liquid water.'

After Earth, the next rock out from the Sun is the Red Planet, Mars. Only half the size of our planet, Mars manages to pack in some spectacular scenery, including the biggest volcano and the most extensive canyon system in the whole Solar System. Planet hunter Geoff Marcy will never forget how he, as a youngster, saw Mars through a small telescope for the first time.

'If you squinted and looked in the eyepiece ever so carefully, you could see the white polar caps. To actually see white polar caps on another planet clearly rendered that world kin to our own Earth. It was completely unforgettable, and begs the kinds of questions that now we're able to answer. You know, there's got to be ice there – some of it's water ice, some of it's carbon dioxide ice as well – so there have got to be places

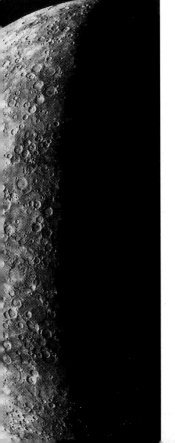
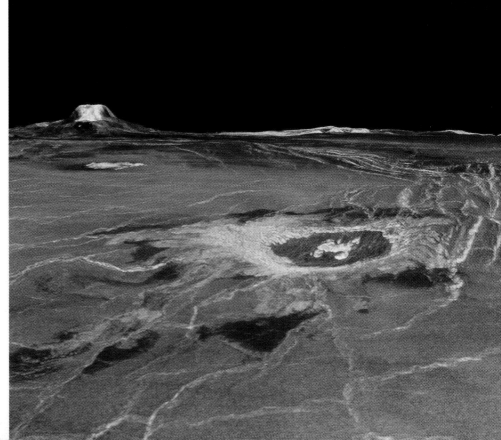

where you can at least ask the question whether or not there was ever liquid water on the surface.

'I'd go to Mars in the blink of an eye – I would absolutely go immediately. I wouldn't care what the chances of surviving were, and I wouldn't care how long it took. I would just go.'

John Spencer wouldn't be in such a hurry to travel to Jupiter, the next planet out. 'It's a really dynamic place. It doesn't have a solid surface, but it has enormous storm systems. The most famous is the Great Red Spot, which may have been seen three hundred years ago and is still going strong. But there are many smaller storm systems, and these can last for decades – utterly different from Earth where a storm might last only a few days or a week or two. The other big difference is that the winds on Earth blow in all kinds of directions. But on Jupiter, they mostly blow just east or west, and spread the clouds out into bands. When you look at Jupiter through a telescope you see these alternating light and dark bands, which are due to clouds being strung out by these very powerful east and west winds.'

Jupiter is so massive that it also controls its own mini solar system. It holds twenty-eight moons in its gravitational thrall – four of which are a match for the planet Mercury. 'The four big ones were discovered four hundred years ago, as soon as Galileo invented the telescope,' recounts John Spencer. 'And these moons are fascinating places – each is totally different from the other. Io is the closest in to

ABOVE LEFT *The Blue Planet, Earth – just the right distance from the Sun to be able to sustain liquid water and abundant life.*

ABOVE CENTRE *Although Mars is only half the size of Earth, it has a canyon system as wide as America and a volcano that would cover the whole of Spain. And it almost certainly harbours primitive life...*

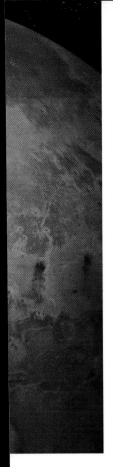

Jupiter, and is covered in erupting volcanoes produced by the gravity of Jupiter affecting the interior of Io. The next moon out is Europa, which has an icy surface – and we think there may be enough heat generated by Jupiter's gravity on Europa to produce an ocean beneath that thin, icy crust.

'Ganymede, still further out, is the biggest moon in the Solar System – it's actually larger than Mercury, though it weighs a lot less because about half of it is made up of ice. It seems to have had a pretty complicated geological history, and there may be some kind of liquid layer beneath its surface, but not to the same extent as Europa. And finally there's Callisto, which is completely covered in impact craters.'

But even mighty Jupiter doesn't win the title of the most multi-mooned planet of the Solar System. That honour falls to Saturn, which has at least thirty satellites in tow. Another gas giant, just a little smaller than Jupiter, Saturn can also boast the most sensational rings of any planet (although all the gas giants have rings circling them).

Beyond Saturn lies Uranus – the first planet to be discovered since antiquity. Barely visible to the unaided eye, Uranus was found by the amateur astronomer and professional musician William Herschel in 1781. It's a bland, boring gas giant – smaller than Jupiter and Saturn – with a family of twenty-one moons.

Neptune, discovered in 1846, is Uranus' twin in size. But that's where the resemblance ends. While Uranus shows virtually no features in its cloud-deck, Neptune – for whatever reason – is stormy and active. When the *Voyager 2* spaceprobe flew

ABOVE *Up close and personal with mighty Jupiter, and its vivid swirling cloud-belts. The Great Red Spot (right) has probably existed for more than three hundred years and is three times the size of the Earth.*

past the planet in 1989, it took stunning images of 'the Great Dark Spot' – a storm the size of Earth – and brilliant white high-altitude clouds.

Bringing up the rear in the Solar System is tiny Pluto, found by Clyde Tombaugh at the Lowell Observatory in 1930. Smaller than the Moon, icy Pluto is considered by some astronomers not to be a genuine planet at all. Since the recent discovery of dozens of tiny 'ice dwarfs' swarming around the outskirts of the Solar System, many scientists have demoted it to the status of 'chief ice dwarf'. Pluto is so far from the Sun that it takes 248 years to trundle around once.

This is a far cry indeed from the first extrasolar planet to be discovered, a world that takes just four days to orbit its star. And it's a world the mass of Jupiter – by now, no means the only one to have been winkled out by the planet hunters. The planets in our Solar System might have wacky personalities, but those orbiting other stars are turning out to be downright weird.

'The discovery of Jupiter-like planets close to stars has really thrown ideas of the

ABOVE *Saturn with its glorious rings – surely the ultimate space-tourism destination of the twenty-second century?*

formation of the Solar System into a tizzy,' admits John Spencer. 'Nobody expected this. We expected planetary systems to be like our own, but we haven't found one yet. And the ones we've found are something that people would have thought was impossible before their discovery.

'Planets like Jupiter have to form where it's cold, because that's the only way they can accumulate the gas they need to get as big as they get – if the gas is too hot, it just won't accumulate. And yet here they are, very close in to their stars, and very warm.

'It's very exciting. I mean, we always love it when our expectations aren't met, and things turn out to be different – that's how you learn more. It's been one of the most exciting discoveries in planetary science since I've been involved in the business – and that's twenty or thirty years.'

The weirdo worlds quickly acquired a name for themselves – literally. 'Many of these planets orbiting nearby stars are twenty times closer to their star than the Earth is to the Sun,' points out New Zealand astronomer Andrew Collier-Cameron. 'That means they're a hundred times closer to their parent star than Jupiter is to the Sun, and they're actually receiving ten thousand times more energy. So, since they're being blow-torched ten thousand times more strongly than Jupiter is, "hot Jupiters" seemed a good name for them.'

Living on a hot Jupiter would not be a bed of roses. 'If we got into the location of one of these hot Jupiters,' warns Seth Shostak of the SETI Institute, 'all I can say is that you should invest in air-conditioning stock, because it's going to be very unpleasant. The oceans would immediately boil away, and everything would be kind of soft and liquid. You wouldn't be able to survive on a planet like that.'

Collier-Cameron agrees. 'I wouldn't care to go there for my summer holidays. The temperature would be about eleven hundred Celsius, and you would actually have clouds made of molten iron droplets. I was talking to some people from British Steel

ABOVE *A moon circling a hot Jupiter would hardly be a bed of roses to live on. In this artist's impression, pools of molten rock seethe and bubble in the cracked landscape.*

the other day, and they were pretty horrified at the concept of boiling steel. It's clearly not something they can do easily, even in their blast furnaces – it does require very high temperatures.

'But it's great fun to imagine what the weather would be like on a hot Jupiter,' he continues. 'For one thing, the planet is so close in that it will always keep the same face towards its star, as the Moon does towards the Earth. So one side of the planet is getting blasted by radiation from its star, but the other side always faces away.

'You can imagine that the weather systems generated on the sunward side are constantly driving winds over to the dark side of the planet, and supersonic breezes would blow hurricane systems across the day–night boundary – where the stuff would sink down on the night side. It's probably more complex than we can imagine, but there's no doubt that the storm systems generated in a place like this would be terrifying.'

Most bizarre of all would be the colours of these hot Jupiters. 'The atmospheres on these planets would mostly consist of a very transparent mixture of hydrogen and helium gas,' explains Collier-Cameron. 'You'd be able to see quite a large distance through the atmosphere, except in yellow light, because there's a small amount of sodium mixed in, and this absorbs yellow. So you'd only really be able to see the red and blue light coming out, which would make them a gentle magenta colour.'

Geoff Marcy, discoverer of many of these hot Jupiters, would love to visit one. 'One of my great dreams in life – and who knows, maybe it'll happen – is to fly a spaceship with some new-fangled propulsion system close to one of these extrasolar planets. And what you'd see at first, of course, is just a bright star. As you got closer and closer, you'd see the planet orbiting the star, and undergoing phases like the Moon. There'd be a crescent planet, a gibbous planet, a full planet and so on.

'Closer still, you'd see meteorology – weather, winds, hurricanes. You'd probably see lightning as well, because these planets almost certainly have magnetic fields like Jupiter and the Earth. These magnetic fields would arc into huge lightning bolts like the ones we see on Jupiter.'

Intrepid Andrew Collier-Cameron even fantasizes about flying in a balloon over a hot Jupiter. 'Imagine what it might be like to go hot-air ballooning on one of these planets. Your balloon would be whipped along by wind speeds probably in excess of several hundred kilometres per hour. You'd be blasted by the star from above. If you were below the cloud-deck, it would be very warm, and you might find rock-droplets falling past you.

'And as the winds carried you round to the night side of the planet, you'd perhaps begin to see some of the dull red glow from the planet's hot interior lighting up the underside of your balloon, through breaks in an iron cloud-deck several thousand kilometres below.

'If there were breaks in the cloud-deck above, you would be able to see out to the clear sky beyond. There, you might see curtains of dancing yellow light where charged particles from the star sweep back in along the planet's magnetic field lines, colliding with sodium atoms in the atmosphere to give you a nice auroral display. The stellar wind on a planet like this would be four hundred times as intense as it is on Earth, so the aurorae would be very, very spectacular.'

Notwithstanding, none of these planets has a right to be there, according to accepted astronomical wisdom. The thought of a Jupiter-like planet up so close and personal to its parent star turned the theory of planetary formation on its head, and left theoreticians having to rewrite their textbooks. 'Bizarrely enough, they actually revived some theories that had been made during the mid-1980s,' recalls Collier-Cameron. 'These suggested that planets like Jupiter could in fact suffer orbital decay and spiral in towards their parent stars. So those theories got dusted off in a hurry.'

Theoretician Doug Lin was one of the first on the scene. 'I recalled some of the work done in the eighties on the formation of giant planets such as Jupiter. We always had a problem holding Jupiters where they formed, because they have a tendency to interact with the disc out of which they were created, and have a tendency to spiral in towards their host star.

'What was surprising in this particular case is not the fact that a planet could form and migrate in – that caused me a great deal of joy. It was that it could migrate all the way in and stop almost next to the surface of the star. And this is where a new theory is required – to account for these short periods.

'Instead of being disheartened by this discovery, I was very excited – excited because of two reasons. One is that some of the ideas we'd been working on for some time had now been confirmed by observation. And the second reason is that the theory we've been working on does not have the complete answer, so this is a new piece of the jigsaw puzzle for us to work on, a new challenge.'

Don Brownlee, an expert on dust particles in the Solar System, remembers Lin's original work. But he also points out that it has a downside – at least, where planets like the Earth are concerned. 'The idea that Jupiter would form out further and then decay in is exciting from an astronomy standpoint. But it's terrible from a terrestrial planet standpoint, because the planets closer to the star, in the habitable zone, would get pushed into the star ahead of it.

'It's death for terrestrial planets. It's the most draconian thing you can think of, having a nice Earth-like planet and having it thrown into the star.'

'If Jupiter had migrated in towards our Sun,' adds John Spencer, 'we wouldn't be here talking about it.'

If migration is common in planetary systems around other stars, could it have happened in our Solar System? Might our familiar nine planets be the last survivors of

a previous family of worlds? Doug Lin doesn't rule it out. 'There's a definite possibility that in our own Solar System there may have been a previous generation of planets that was formed before our own. Those planets would have been born and then migrated into our Sun.

'In fact, what we see in our Solar System may be the survivors of several generations of proto-planets. As they form, they migrate, and undergo infant mortality. So the planets we see in the Solar System today may be the last of the Mohicans.'

If migration happens, it seems to take place very quickly. When Dave Charbonneau measured the size of the planet that made a transit across the face of the star HD 209458, he was surprised that, although it was the same mass as Jupiter, it was very much larger. 'The theorists put their heads together and said, "Does the size that these guys have measured for this planet make sense?" And the answer was no. If you take Jupiter in our own Solar System and move it in close to the Sun, it puffs up only a tiny bit – nowhere near what we found this planet to be.

'So the theorists set about trying to explain how this planet could be about forty per cent larger than our own Jupiter. And what they found was that the planet had to have moved in towards its star very early on in its life – about ten million years after forming. Now ten million years is an awfully long time for you and me, but on astronomical timescales it's a very, very short period of time.'

Theory and observation in astronomy are supposed to go hand in hand – but that requires an ideal world. Not all observational astronomers are great fans of the

ABOVE *Birth of a new planetary system: a computer simulation of the formation of planets around a young star, in which discs of debris left over from starbirth are about to create new worlds.*

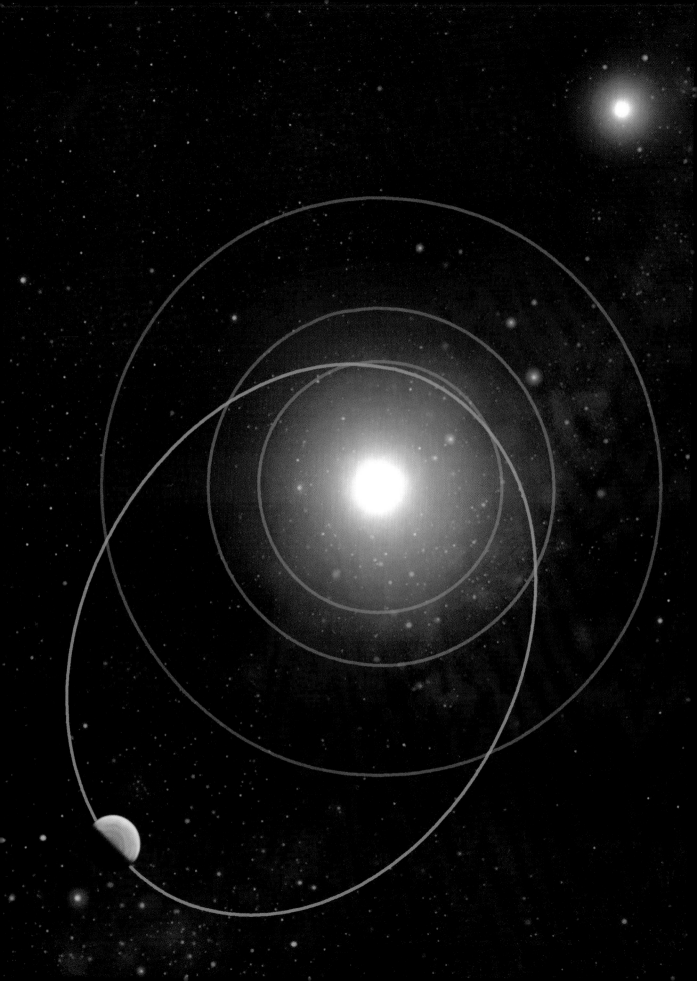

theoreticians, or believe all their predictions. Seth Shostak has some characteristically pithy remarks on the latest ideas. 'Just because you have big planets in close, does that mean that the little guys which you can't detect aren't out there? Well, we don't think so.

'On the other hand, it's been pointed out by the theoreticians – who are always much cleverer than the experimentalists – that when the big planets spiral in they sweep all the small planets out of those systems. Mind you, that's a theoretical argument. And one thing you know if you do astronomy for very long is that theory is always subject to revision after you've used a telescope on the system.'

'I have a love–hate relationship with the theorists,' echoes Geoff Marcy. 'They don't go to the telescope, they don't take data, they don't bother with facts. They just sit there with a pencil and paper – and OK, maybe a computer – and they make things up. And the reason it's a love–hate relationship is that the theorists are the ones who told us that all of the Jupiters would be orbiting far from their star – they couldn't have got it more wrong.

'On the other hand, they're building new theories, armed with the new data we're providing about actual planets going around stars. They've gone to the blackboard, they've erased the whole thing, and they've started over with new equations, new physics, trying to work out how planetary systems form.

'Fifteen different theorists will come up with thirty different ideas. Maybe one of them is right. But of course, that's the way science proceeds. We're a sort of self-critical organization – we do our best and the theorists do their best, and in the end we hope that one theory will win out and prove to be the correct one.'

Doug Lin isn't too fazed by comments like this; he's having too much fun being a theoretician. 'We're working at maximum capacity to keep up with the observers, and as they bring us more and more data, we're developing more and more sophisticated theory. This is a theorist's heaven. In fact, it is a golden age of planetary astronomy, in the sense that observation and theory are developing hand in hand, and the technology is here for us to make these important discoveries. Every month, there's something new.'

So how does he explain how hot Jupiters spiral in towards their star without plunging into its fiery embrace altogether? At the moment, the theory is still being worked on – but here's an interim report.

'Well, the planet's migration is a bit like a motorcyclist on the wall of death,' explains Lin. 'At the top it moves relatively fast, and as it gets slowed down it gradually moves downwards. If the motorcyclist can then give a little bit of extra throttle before reaching the bottom, he or she can maintain that level. And as a planet loses spin and gradually spirals in towards its star, at the very end the star gives it a little extra kick, enabling the planet to maintain its very close orbit.'

As more and more extrasolar planets were discovered, it rapidly became apparent

OPPOSITE *Doomed planet: astronomers have now found over thirty planets in wildly elliptical orbits where they're alternately frozen and baked. They're very unlikely to be abodes of life.*

that the observers were due to inflict even tougher demands on the theorists. Paul Butler, Geoff Marcy's partner in crime, sums up the situation. 'Prior to 1995, the dominant paradigm for planetary systems was that they would be like our own. You'd have the planets in these beautiful nested circular orbits, with small rocky planets like Earth and Mars close to the star, and gas giants like Jupiter and Saturn further out.

'And that vision was basically shattered by early 1996. We found six or seven planets very quickly, and the planets were either these 51 Peg planets close in to their star, or these elliptical eccentric planets.'

What the planet hunters were increasingly finding was that, far from being in well-behaved circular orbits – as we find in the Solar System – planets of other stars had elongated oval paths that subjected them to violent temperature extremes. 'The implications are profound,' notes Butler. 'You wouldn't expect to find life in an eccentric system, because for part of its orbit the planet swings way too close to its star and boils, and for the other part of the orbit it swings too far away, and everything freezes out. It would be a tough game for evolution to get a toehold.'

'We found a planet around the star 70 Virginis that was a real weirdo,' adds Geoff Marcy. 'The orbit was so elongated that people thought to themselves, "Is this really a planet or is it some other kind of beast? Maybe it's a failed star or a brown dwarf, or some other as yet unanticipated type of orbiting object."

'But sure enough, it turns out it's a planet, and we've found some thirty or forty more of these planets in elongated orbits. And we now realize to our surprise that most extrasolar planets are not in circular orbits but in these elliptical ones, and we don't have a good explanation for it at the present time.'

Aspiring hot-Jupiter balloonist Andrew Collier-Cameron imagines being there: 'If you put Earth in an orbit that, say, took it from the orbit of Mars during one part of the year and then in towards the orbit of Venus, you'd have a very vicious winter when the Earth was furthest from the Sun. And at closest approach, you'd probably find that the oceans would begin to boil. It would be a very unfriendly environment.'

The puzzled observers turned to theorist Doug Lin for an explanation. 'The latest discoveries suggest that most of these planets with elliptical orbits have signs of additional planets,' he observes. 'Whenever planets form, they tend to form in a family. And, as you know in a family, siblings tend to perturb each other. And sometimes this interaction becomes so strong that it can break up the family.

'Likewise, in these planetary systems the interaction between planets can cause the orbits to become ellipses. Sometimes it can be so totally disruptive that some of the planets may even get kicked out from the neighbourhood of their host star.'

Are the stable, circular orbits of the planets in our Solar System some kind of freak occurrence? Is the Solar System the exception rather than the rule? 'The more massive the planets are, the more intensely they interact with each other, and the more likely that they will disrupt each other's orbit,' explains Lin. 'In our own Solar System, we are atypical because most of the planets are relatively low in mass.

'Now is it really that we are atypical?' he muses. 'My personal feeling is that we are not, because with the current technology we are finding systems that are the easiest to detect – with planets that are relatively massive. As we perfect our observational technique, we'll find planets with a lower mass. These systems will be very stable – they'll resemble our Solar System much more. So maybe the first gold nuggets we are picking up today are the oddballs, and the true nature of planetary systems out there in the Universe is yet to be discovered.'

But Doug Lin warns that the cosy stability of our Solar System may not last for ever. Our planetary system is relatively young – the planets are in paths little different from when they were formed. But as Lin cautions, 'At different ages, they have different personalities. Many extrasolar planetary systems have reached the stage when they have gone through divorces, they've been through battles – and what they are left with is the result of the battle scar.

'When you see these systems with highly eccentric orbits – these systems so different from our own – perhaps that represents the destiny of our own Solar System.'

So are planets ultimately bad news for one another? Are they either massive brutes which steamroller their smaller siblings into their host star, or unruly orbs that throw their neighbours into the limbo of an elongated orbit – or out of the system altogether?

'Well, to say a planet is good or bad is to give it a moral attribute that the poor planet doesn't deserve,' laughs Paul Butler. 'I mean, it's just a planet. You have to realize that, prior to 1995, there were no known planets outside of the Solar System. So we're acting like an umpire or referee. We don't care if things are good or bad – we just take the data and tell people what we've found.

'The 51 Peg planets are telling us that planets can form far out, but then can move very close in to their star. Similarly, the eccentric planets are telling us that planets interact, and their orbits can be thrown. So we're learning a tremendous amount by the discovery of these new objects, and it's helping to put our own Solar System in a new light.

'For instance, we now realize that if Saturn were about twice as massive, it would gravitationally interact with Jupiter on a short timescale. One of them would get thrown in, and the other one would get thrown out – and we wouldn't be here having this conversation.'

For planetary scientists like John Spencer, the discovery of these weird new worlds has opened a whole new Aladdin's cave to explore and wonder at. 'Given all the strange phenomena that we find in our own Solar System, we can only begin to imagine how strange things might be elsewhere in the Universe. We're going to find amazing stuff out there in those other solar systems – but we don't know what it is yet.'

Signs

chapter three

that we discover will undoubtedly be far beyond our own

# of Life

At NASA's Ames Research Center in Silicon Valley, biologist Lynn Rothschild contemplates her lot. 'I think I have the best job on planet Earth,' she enthuses. 'Because I love microbes. I love to find out what they're doing. And to do this, I need to go to extreme environments that keep out things like fish and trees. I've worked in places where the temperature has been very high, and there are all sorts of microbial communities that are just fabulous colours. I've spent a lot of time working in really acidic regions – more acidic than lemon juice, or even more than that.

'And I've worked in high salt conditions where you have piles of salt with microbes in them that are bright red. You feel you're in another time and another place. It's better than anything you could think of for science fiction. You get to journey back billions of years through Earth's history. You journey out into the Solar System, and you can imagine what it's like on other planets, and even beyond on planets that circle other stars.'

Life. That's the driver that motivates people to search for planets around other stars. They're looking for a planetary system like our own, without bulldozing hot Jupiters or worlds in wildly eccentric orbits that get alternately fried and frozen. Most of all, the planet hunters are looking for an Earth out there. A small warm world, with the potential to develop life – but what kind of life?

And at the same time, the community of bioastronomers is trying to achieve the unimaginable: if there *is* life out there, might it be more sophisticated than microbes? Could it be intelligent? Might we be able to contact it?

'I think the big question for every human being is, are we alone or is there other intelligent life in the Universe?' muses University of California theoretician Doug Lin. 'If there are ETs, where are they and how do we recognize them? Now, you don't find ET in one day – just as Rome wasn't built in one day. You build Rome brick by brick, building by building, street by street and complex by complex. That's how a city emerges.

'You have to find systems around stars which may have planets, and find out whether those planets may be suitable for life to form.'

'We don't know really what fraction of stars are going to have good planets,' admits Seth Shostak of the SETI Institute. The Institute – housed in a breezy minimalist Californian barn of a building in the heart of Silicon Valley – is the nerve-centre for the Search for Extra-Terrestrial Intelligence. Perfectly sober, highly qualified scientists spend their days thinking about the nature of aliens, and how we might get in touch with them.

'We can only guess at the number of good worlds where ET would be happy, and evolve into something technological,' continues Shostak. 'We're talking about the kind of worlds that would cook up a little bit of interesting dirty chemistry we like to call biology. It may be that the correct number is one or two per cent. But it could be that, of the ninety-eight per cent of stars that don't have giant planets, maybe they just have small planets. Perhaps every star is a good candidate. We really don't know that yet.'

OPPOSITE *The variety and profusion of life on Earth is astonishing. But is life as plentiful and varied elsewhere in the Universe?*

And there are plenty of stars at our disposal, as Shostak points out. 'If you had to count the stars in our Galaxy one by one – by the way, I don't recommend this exercise unless you're having difficulty sleeping – and say you counted them at the rate of one a second, it would take you well on the order of ten thousand years.'

But the bottom line is understanding how life gets started in the first place – and no one yet has an explanation for that. What is it that makes the difference between a mass of gooey chemicals and a sentient being? The question is probably the biggest imponderable in the whole of science.

Seth Shostak sets out the basic parameters. 'In order for life to form on a planet, the requirements are fairly straightforward. You need a source of energy – most

ABOVE *The raw materials of life are scattered throughout the Cosmos. Giant nebulae are rich in our very own building blocks – the carbon in our bodies, the oxygen we breathe, and even the constituents of water.*

planets will have that from their own sun. Beyond that, you need a little bit of chemistry to get going, and for that we suspect you need liquid water. Liquid water is considered the sine qua non of life – if you have liquid water, you might have biology.'

'There have been experiments that have been done to show that the building blocks of life – certainly amino acids – can be formed literally overnight,' explains Lynn Rothschild. 'What we don't know is how long it takes to go from these very simple compounds all the way to a fully-fledged organism.

'Now, going from an organism to something we would recognize, like an elephant, or a human being, took quite a period of time on Earth – several billion years. I'm not convinced that it would take that long if we put the clock back and ran that sort of experiment on another planet. That being said, I don't put my cultures of microbes to bed at night and worry that I'm going to see Frankenstein the next morning.'

Life took root on Earth very early on in its history. A billion years after the Earth was born – when it was just one-fifth of its present age – the first single-celled creatures began to appear. But forming life, as Lynn Rothschild points out, does not necessarily guarantee swift evolution to higher lifeforms. Many biologists are of the opinion that life will appear on a planet which has the right conditions, but will never progress beyond the phase of being 'green slime'. And even if it attempts to, it will find that there are obstacles in the way.

'It seems to me that the conditions that allow the formation of life, and then the conditions that allow life to survive aren't necessarily the same,' observes geologist Peter Ward. 'You start out with conditions that allow life to get a foothold, but then you have catastrophes. People assume that once you get to life, it's going to inevitably evolve to complexity, inevitably evolve to intelligence. But do you have enough time? Catastrophes occur in the Galaxy, and they've happened on this Earth – and they happen a lot more frequently than we think on planets other than our own.'

The processes that create planets also create danger zones for burgeoning new worlds. Planet-making is a messy, violent operation: huge rocks and boulders collide and dash one another to pieces, until they can build up enough mass to make the beginnings of a new world. Even after the work of formation is over, the rubble in the building site remains, smashing into the young worlds and wiping out most of the murmurs of incipient life. But the rule of survival of the fittest holds true. Impacts from space have probably driven evolution on Planet Earth.

Sixty-five million years ago, a ten-kilometre-wide rock from space hit the Yucatán Peninsula in Central America. The resulting pall of dust, which drowned out the sunlight for nearly a year – not to mention vicious forest fires and huge tidal waves – wiped out most of the animal species on Earth, including the dinosaurs. But small furry mammals, which could shelter in holes in the ground and hibernate when food was scarce, managed to eke out an existence. They became the dominant creatures on the planet – and are our direct ancestors.

'If the dinosaurs hadn't been wiped out sixty-five million years ago by an asteroid or comet impact,' attests Don Brownlee, 'we wouldn't be standing here. It would still be the age of the reptiles.'

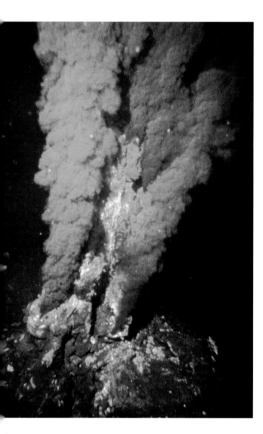

Even if impacts have a totally devastating effect on higher lifeforms, microbial life can hang on in there. It's made of sterner stuff.

'Let's say we're hit by a giant asteroid,' ventures Peter Ward. 'It cleans out the complex life – which has a narrow range of habitability and living conditions – and leaves the bacteria behind. It may be that planets become sterilized with their animals and plant equivalents, but the bacteria and microbes get left behind. I like to think that once microbial life is on a planet, it has infected it – and, as we know, it's hard to get rid of infections.'

And one of the major discoveries of the past few years is that bacteria and microbes are even more tenacious than we ever imagined. They have been discovered in environments where conventional biology dictated that life couldn't survive – in deep-sea vents where superheated gases from inside the Earth make the oceans boil; down bottomless boreholes; encased in rocks in the frozen wastes of Antarctica; and in the acid pools surrounding seething hot springs.

Lynn Rothschild has been part of this incredible revolution. 'During the last twenty or thirty years,' she explains, 'we've completely expanded our envelope of where we think that life can survive. Thirty years ago, we wouldn't have thought that organisms could live at temperatures of boiling water or in extremely acidic or alkaline conditions – or in extremely solid salt, or with no water. There are organisms that live down at the bottom of the sea under incredibly high pressures that cannot deal with the kinds of pressures we deal with every day.

'And we now know that, not only can organisms live under these conditions, but they actually thrive there. We call them extremophiles. To some extent, what we consider extreme is from our perspective. We regard a nice pleasant day in May in London to be the norm, but there are many organisms that would consider conditions like that extreme. When you think about it, there are very high concentrations of oxygen, which would be toxic to many microbes.

ABOVE *'Black smokers' – hydrothermal vents submerged deep down on the ocean floor – support communities of living creatures that need no sunlight or air to survive.*

OPPOSITE *Yellowstone Park's boiling-hot springs teem with micro-organisms that can live in extreme concentrations of acid. These extremophiles give scientists confidence that life has a wider range of tolerances than previously thought.*

'One of the places we've found life in the last ten or fifteen years is deep underground. It turns out there's a very active colony of tiny little worms – nematode worms – that are perfectly happy beneath the surface. In fact, organisms have been found literally kilometres beneath the surface of Earth, which suddenly expands the whole habitable area of Earth greatly. And this expands the possible places where life could be found on other planets.'

Rothschild is especially drawn to microbes living in hot springs. She has worked in many extreme environments, but Yellowstone Park in Wyoming keeps her coming back. She's a brave woman. As she takes samples from an acid pool here, she's aware that she's dicing with her life. If you fall in, you can opt to be boiled, or killed by arsenic poisoning a few days later. But it doesn't faze her.

'Yellowstone is a wonderful place to work,' she maintains. 'It really is like being in a time capsule. You can go back to what the Earth was like three, four billion years ago. I come here to be in that time capsule and to see how micro-organisms – particularly algae, which are related to the seaweeds you see on the beach – really go about their day.

'Most of the extremes that we think about would be absolutely impossible for humans to live in. We can't stand the sort of acidity that a micro-organism can stand. I have microbes that I work on that I've put in sulphuric acid – that's like battery acid – and they've survived perfectly well for an hour. Take 'em out, and they're fine.'

These new findings about the tenaciousness of microbial life, and the truly amazing extremes that it can tolerate, led Rothschild to check out if it could survive the ultimate frontier: space itself. 'I know this sounds like a totally ludicrous idea, but our group – as well as others in Europe – has taken micro-organisms, put them on a satellite, and exposed them to the space environment.

'The space environment is really nasty. It's incredibly cold, it's a huge vacuum, incredibly dry, and the radiation levels are very high – both in ultraviolet radiation and charged particles coming from the Sun and from outer space.

'We expose the organisms, bring them back to Earth, and look at them in our lab. Some organisms can survive weeks, some even months. So it's not so crazy to think that life could go from planet to planet.'

ABOVE *The desolate wastes of Antarctica are home to abundant microbial life – it even thrives inside rocks.*

The idea that life on Earth might have arrived from space is not new. Ironically, the cosmic harbingers of doom – asteroids and comets – are also believed to harbour the rich chemical mix that may have seeded the Earth with life and water in the first place. Rothschild's research also raises hopes that primitive life might be able to travel space independently. 'We know how it can get there on an asteroid or comet,' she notes, 'and now we know it's possible for microbes to survive for at least a couple of weeks or months, or maybe longer.'

So – might other worlds in the Solar System be 'infected' with life? 'Right now, we're focusing on Mars and Europa, one of the moons of Jupiter,' explains Rothschild. 'Europa because we think that today it has a liquid water ocean underneath ice, just like the lakes in Antarctica that have ice on the surface and liquid water underneath – and yet there are organisms that live there.

'On Mars we have good evidence there was liquid water in the past – there isn't any today on the surface – but now we're starting to rethink that. There may be water just below the surface, or perhaps even a kilometre down. And we know there's life a kilometre beneath the surface of the Earth, so perhaps there's life beneath the surface of Mars.

'I think right now these are our two best candidates. We know we can get there, we know what we're looking for, and we know something about the liquid water conditions. So I'm extremely excited about it.'

Don Brownlee shares Rothschild's enthusiasm. 'One of the most exciting things I can think of is looking for life in the Solar System. Because here, we can fly spacecraft to it, investigate it, and bring samples back to Earth where we can study it in phenomenal detail.

'And there's a reasonable likelihood that we will find either microbial life or evidence for past microbial life on places like Europa and Ganymede and Mars. And it's a wonderful opportunity – I mean, it's the first time in history we've been able to have this capability.'

Europa also fires the imagination of moon expert John Spencer. 'It's going to be hard to find life because it's below this surface layer of ice, and we're going to have to drill down. But we're still very excited about the possibility.

'If there is life on Europa in this ocean beneath the ice, it could be something like what we find around hydrothermal vents on the Earth's ocean floor. That's where chemical energy from volcanic activity is coming up to the surface, and is providing energy for the little communities that exist around these vents. On Europa, we don't expect that sunlight would get through the ice shell, so any life down there would have to depend on something like deep-sea vents.'

'I think one of the great surprises of the last decade has been the growing realization that a place as obviously inhospitable as one of Jupiter's moons might in fact harbour living organisms,' adds Seth Shostak. 'There may be Europan tuna or something like that living beneath the ice.

'Now, Jupiter's five times as far from the Sun as Earth is, and that means that the sunlight is – what? – four per cent of what we get here. And that might be our nearest cosmic company – albeit fairly primitive company. I think that's a big revelation. It shows that biology may be inhabiting places that we had never before considered to be very likely worlds for life.'

But Shostak concedes that life might just lie closer to home – on Mars. 'At the turn of the last century, there was this whole business of the canals on Mars. People were claiming that they saw thin lines crossing the surface of our little ruddy buddy there. In fact, by the 1920s, very few scientists believed that was true. But even so, at around that time, attempts were made to try and listen to radio signals from Mars in case there were sophisticated beings there broadcasting evidence of their presence.'

The whole canals episode was stoked up by Percival Lowell, a rich Bostonian banker and enthusiastic amateur astronomer. In 1877, he heard that the Italian astronomer Giovanni Schiaparelli had reported seeing 'canali' on Mars. The word actually translates to 'channels', but Lowell – who could also see the features through his telescopes – believed that they were artificial canals, and the artefact of an intelligent race. So obsessed was Lowell with his canals that he built a sophisticated observatory amidst the pinewoods of Flagstaff, Arizona, to investigate them in more detail.

Lowell's imagination knew no bounds. He devoted his life to mapping the canals, seeing changes as they appeared to shrink, grow or become double. He published books that fired the public's imagination, explaining how the Martians were clinging on for

ABOVE *The cracked, icy surface of Jupiter's moon, Europa, may conceal a deep warm ocean inhabited by weird alien lifeforms.*

'So the temperature extremes on a hot Jupiter may not be as critical as the question of water. The other question is how stable the climate is. If the climate shifts a bit, we have serious problems with life as we know it. The organisms that live at incredibly high temperatures are not the same ones that can deal with the freezing temperature of water, and vice versa.'

The other downside to any organism living on a hot Jupiter is radiation damage. If a hot Jupiter is built anything like its counterpart in the Solar System, it will have a rapidly spinning liquid core generating a powerful magnetic field – which means a searing radiation environment. But it seems that even this mightn't rule out life altogether.

'In the last five years of my own research,' explains Lynn Rothschild, 'I've tracked these little algae to see how the Sun – and particularly the ultraviolet portion of the solar spectrum – affects their genetic material.

'The results have been absolutely astonishing. If we're right, what we're showing is the ultraviolet damage to the DNA. They suffer enormous damage during the day. But then they spend maybe fifty per cent of the energy that they would normally have spent on DNA synthesis actually repairing the DNA.'

Hot Jupiters aside, what about finding life on the other worlds in planetary systems beyond our own – especially those in wayward, eccentric orbits? 'You certainly can't have life-bearing planets whizzing around in eccentric orbits,' maintains John Spencer. 'You can't have planets migrating in from very far out to very close in, and sweeping up planets like the Earth along the way.'

Peter Ward agrees. 'I think bacterial lifeforms are easy to produce, and Earth's record tells us this. We had life on this planet almost as soon as we physically could have. But then, why did it take another three to four billion years to get the simplest type of animal, and then much longer after that to get intelligence?

'You have to have long periods of stability, and most places in the Universe won't give you that. I think lifeforms generally would be very common, and that complex lifeforms would form just as easily as simple ones. But it's that stability you can't find.'

But Lynn Rothschild believes that *instability* might be just what life needs to be given a helping hand in the beginning. 'In evolutionary biology, we've believed for years that we need some stability for evolution to occur, including the evolution of life. I'm not convinced that we really need that. It may be that having a fluctuating environment, having rapid changes in temperature, pressure or lightning coming in might be just what we need for creating life quickly.'

Everyone, however, is agreed about the vital ingredient needed to create and maintain a stable planetary system, with well-behaved near-circular orbits and no gravitational hanky-panky. It's a Jupiter – but not a hot one.

'There can be good Jupiters and bad Jupiters,' explains Don Brownlee. 'Our

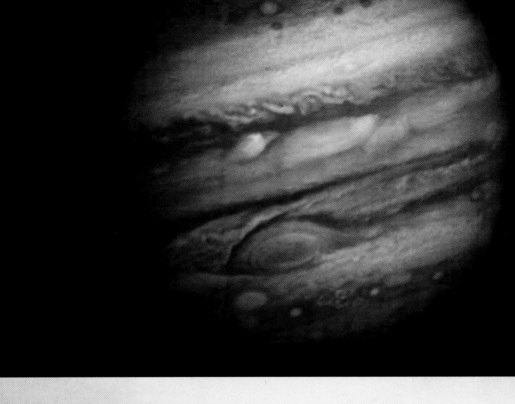

Jupiter's a good Jupiter. It diminishes the impact of dinosaur-killing objects on the Earth. But there are also bad Jupiters. I mean, a bad Jupiter could fling in too much material or push a planet like ours into the Sun.'

John Spencer elaborates. 'We think a planetary system with a planet something like Jupiter – at about Jupiter's distance – is going to be much more conducive to life. Maybe we're just being self-centred in this respect, but if we can find a system with a cold Jupiter – a Jupiter-like Jupiter – then that will be the kind of place where we'll think that life is more likely.

'Jupiter is well over half the total mass of the Solar System. It really controls everything that goes on in the planets. It was important when the Solar System was forming, and it stood up for itself by its own gravity – it prevented a planet forming where we now have the asteroid belt. Its gravity also controlled the orbits of the other planets.'

Geoff Marcy echoes Spencer's sentiments. 'The real interesting goal for planet hunters right now is to find a Jupiter-like planet that orbits as far from its star as Jupiter orbits. Finding a Jupiter would be a signpost of the existence of the smaller terrestrial planets that we can't yet detect.'

As well as providing stability in a planetary system, cold Jupiters have an equally important role to play. 'Jupiter has a critical role for life on Earth,' acknowledges Marcy. 'Many people don't realize that it gobbles up the comets and asteroids which would be

ABOVE *Our own Jupiter is a 'good' Jupiter. Its gravity protects the inner planets, like Earth, from bombardment by asteroids and comets.*

OPPOSITE *A cold Jupiter, as seen from its watery moon in this artist's impression. The moons of extrasolar planets could well be abodes of life.*

swirling around the system all the time. We all know that occasionally comets do hit the Earth, and make big craters.

'But Jupiter, over these billions of years, has sucked up most of those comets and asteroids – the debris of our Solar System – thereby protecting us from the impacts that would be coming from all directions. So we're actually very lucky that we have Jupiter as a Solar System protector, keeping the Earth safe from most of the asteroids and comets.'

Not all comets and asteroids hell-bent on collision with Earth get gobbled up by Jupiter. In fact, Jupiter can act in reverse – as a cosmic sling. 'It's encounters with Jupiter that can steer things towards the Earth or deflect them away,' observes John Spencer. And it's a fragile balance between good impacts and bad impacts, as he notes. 'These little objects – they're not very big, but they're very important for us here on Earth in both good and bad ways.

'The bad way that comets and asteroids are important is that if one of them smashes into Earth, it's bad news for life. But the good way is that comets and asteroids have water inside them, and when they smashed into the Earth early in its history, they would have brought water and the other elements essential for life. Otherwise the Earth would have been a dry rock.'

Although astronomers are agreed that a cold, 'good' Jupiter in a planetary system beyond our own would be good news for finding life-bearing planets like the Earth, no such beast has yet come to light. As Peter Ward points out, 'We've never yet seen a good Jupiter other than our own. Every single Jupiter that we have seen so far is a bad Jupiter.'

But veteran planet hunter Geoff Marcy isn't despondent – and actually believes that he may have something up his sleeve. 'What we see in our data is very promising,' he enthuses. 'We see stars wobbling as if they possibly have a Jupiter analogue – a Jupiter that would take twelve years, like our Jupiter does, to go around its star. We've only been working with the Keck Telescope for five years, so we haven't seen one full wobble of a star – in fact, we haven't seen half a wobble yet. But we're seeing hints.'

The planet hunters are also aware that they need to hone their techniques to get the maximum information out of their data. Knowing that the planets are out there, from the gravitational tug they exert on their parent star, isn't enough. 'Finding out what they're really like,' muses John Spencer, 'we need to study the light from those planets directly. And that's hard, because a planet is very small and very faint, and the star is very close to it and very bright.'

But they're working on it. A new instrument being built for the twin Keck Telescopes on Hawaii combines their awesome light-gathering power in a novel way. 'The Keck Interferometer,' explains Spencer, 'will be the first step in getting to the extreme level of detail you need to separate this tiny spark of light from a planet from the very bright star next to it.'

'The new Keck Interferometer is a spectacular machine,' enthuses Geoff Marcy. 'It's one of those things I've dreamed about for decades and now here it is. It will allow us to actually get a picture of a planet going around another star – and we'll probably target young stars, because the young planets going around such a star are still glowing like embers in a fireplace.'

Marcy is looking to the future in more ways than one. 'Right now, we're only able to detect Jupiter-like planets around other stars. Next we'll find Saturns – we're already finding a few – then maybe smaller planets like Neptune. Then, when we're really good at this game, we'll find Earth-like planets.'

Finding an Earth is the Holy Grail of a planet hunter. Andrew Collier-Cameron is in absolutely no doubt as to what he really wants to discover: 'Earths. We have really good prospects of being able to detect Earths with the space missions that are currently on the drawing board within the next few years. If we could find an Earth-size planet with an atmosphere that contained oxygen, we could be certain that life has evolved around stars other than the Sun somewhere else in the Universe. Might not be very sophisticated life, but that would make a pretty good finale to anyone's career in this business.'

'Call me Pollyanna,' adds Seth Shostak, 'but I'm still optimistic that once we have the right instrumentation we're going to start finding Earth-like planets – suggesting that the kind of habitat we have here is really not such a rare one.'

The only way we are going to find Earth-like planets around other stars is from the cold, clear vantage-point of space – like the Kepler Project. Dave Charbonneau,

ABOVE *Earth – the only place where we definitely know that life exists. To find an 'Earth' circling another star is the Holy Grail of planet-hunters.*

whose small telescope in Colorado detected the puffed-up planet passing in front of the star HD 209458, is a huge fan. He explains: 'It's a scaled-up version of STARE,' the Stellar Astrophysics and Research on Exoplanets, 'and it will be in space. The goal is to find Earth-like planets around Sun-like stars. The trouble is that the Earth only blocks a tiny area of the Sun and it takes a year to go round.

'It's a very difficult measurement. But the proposal is to take a moderate-size telescope, put it in space where it can stare at a patch of stars. Then it can basically click away, taking a picture every few minutes for about four years. And the idea is that, for a number of those stars, you're going to see these little blips maybe once a year, when an Earth-like planet passes in front.'

Even more visionary schemes to find alien Earths are on the drawing board. 'There is this ultimate European space mission called Darwin,' explains Didier Queloz, discoverer of the first extrasolar planet. 'The big goal is to see an Earth-like planet and to analyse its light. We want to see if we can detect oxygen or water – you know, the kind of holy elements that are linked to life.'

And European and American astronomers are working together on the Terrestrial Planet Finder. This ambitious mission involves a whole fleet of telescopes sailing together in space. 'Its main goal would be to take a spectrum of the colours from another planet,' explains Geoff Marcy, 'and that will tell us about the composition of the planet, maybe giving us indications whether there's life there. Things like methane and ozone together that are signs of life here on Earth.'

But not everybody is confident that the discovery of an Earth-analogue going around another star would be the answer to our question of whether or not we are alone in the Universe. 'Oh, I think terrestrial planets will be found fairly quickly,' maintains Peter Ward. 'But simply because you have a planet the size of Earth or even the same distance from the Sun, it doesn't mean it has the same chemical composition of Earth. It doesn't mean you'll have oceans.

'Fresh water may be the rarest thing around. And it certainly doesn't mean that once life evolves, it is going to go to complexity. And if you have a meteor

EXTREME UNIVERSE

ABOVE *A fleet of spaceprobes gangs up together in Europe's proposed Darwin mission. Its goal will be to track down 'Earths' around other stars.*

bombardment that's a hundred or a thousand times what it is now, you're not going to have complex life. Without Jupiter – without something to protect you – you're in really deep trouble.'

Don Brownlee echoes his pessimism. 'The newly found planetary systems tell us that Nature is very complicated, and that all planetary systems aren't like ours. We see a huge diversity. This says to me that we live in a magical place – a very special system. And we live on a very special planet in a very special system.'

'We're a planet that has geological conditions that may be actually hard to form,' adds Peter Ward. 'I think if you look at it from a geologist's point of view, we're not commonplace. We're very rare.'

SETI Institute's Seth Shostak begs to disagree. 'There are people who would argue that the particular circumstances of our Solar System – and for that matter, our Sun and our Earth – are very exceptional. That we're just lucky to be here because the slightest thing was different, otherwise we wouldn't be having this conversation. Well, I don't subscribe to that.

'I think the best thing you could do to prove that we're not so exceptional is simply to find evidence of another intelligent civilization elsewhere.'

'The discovery of simple single-celled creatures living elsewhere in the Universe would be fabulous,' agrees Paul Butler. 'But on the other hand, these single-celled creatures aren't going to give us new music; they're not going to give us any new Shakespeares; they aren't the little green men who are going to beam something fabulous to us. They're not the creatures of a Steven Spielberg movie.'

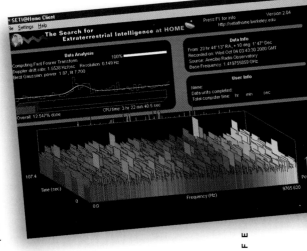

And that's the ultimate goal in the quest for planets around other stars: to discover intelligent life out there. The name of the game is SETI – the Search for Extra-Terrestrial Intelligence.

'The best chance of finding life,' continues Butler, 'is the SETI project. Since we can't travel between the stars, and presumably other life can't, the only way to contact them is with radio waves or with laser beams – some sort of electromagnetic radiation that travels at the speed of light. So at the moment, there are about half a dozen groups in the world scanning the skies day and night for these alien radio transmissions.

'The possibility that SETI is going to discover anything is small. On the other hand, if SETI finds an alien with a radio telescope, it will probably be the single greatest discovery in the history of humanity – greater than fire in terms of its implications.'

ABOVE *Discover alien life in your front room! This SETI-at-Home computer screensaver analyses data from the Arecibo radio telescope to detect an alien signal.*

But how likely is it that green slime – if indeed it does exist on planets around other stars – will develop into intelligent life that wants to broadcast its presence to the Universe? 'If you visited Earth two million years ago,' ventures Seth Shostak, 'you would have found that the smartest critters on this planet were a certain kind of dolphin. And you say, well, these dolphins – they may have evolved to the point where they're constructing radio transmitters.

'Maybe they would, or maybe they wouldn't. All I'm saying is that Nature has experimented with intelligence at various levels in the past. It sounds like Nature does find intelligence a worthy property of its progeny. Perhaps it's an inevitable consequence of evolution. But we really don't know that, and the only way to answer this question is to go find it somewhere else.'

What image does Seth Shostak have in his mind when he conducts his searches for extraterrestrial life? 'I certainly had an image in my head when I was a youngster of what the aliens would be like, because I happen to have grown up in an era when cheap sci-fi was filling the local cinema screens. Every weekend, I would go down with my buddy to see yet another creature terrorize our planet and abduct us for unauthorized experiments.

'I think in those days I expected that the aliens were going to be very much the way Hollywood portrayed them – which is to say, very much like us. And I think that the reason they looked like us in those days is because they had to fit into a rubber suit.

'I see a very direct connection between these films and my later study of the subject. There wasn't much science in them, but there was the emotional content of what it would be like to find other beings on other worlds.'

Perhaps inspired by a diet of sci-fi movies, the groups of researchers engaged in SETI are experts in the field of second-guessing how their unknown quarry might conduct its manoeuvres. They've built container-sized listening-in devices capable of tuning into millions of potential 'extraterrestrial radio stations' simultaneously; now they're looking into the prospect of communicating with ET with laser beams. But is SETI using the right technology?

'There's always the possibility that we're barking up the wrong tree,' acknowledges Shostak, 'looking for radio signals or flashing lasers. These are the kinds of signalling technologies we know about – we've optimized the physics we know. Clearly, there's physics we don't know about, and maybe some of that physics includes ways of signalling that would find better bets than what we're doing.

'I mean, maybe a thousand years ago you asked me how we could get in touch with the putative inhabitants of the Moon, and I would have said, "Well, we'll use smoke signals." Maybe we're in that situation, and one always has to keep that in mind.'

There's also a very real problem in that the Sun is a youngish star, born when the Universe was already several billion years old. Other stars would have had their families by then; intelligent life, if it developed, would have had ample opportunity to steal a march on us. So how would we recognize an advanced alien civilization?

'Any intelligence that we discover will undoubtedly be far beyond our own,' agrees Seth Shostak. 'Clearly, we're not going to find any societies that are less advanced than ours, because they won't be producing the kind of radio noise or light signals that would allow us to detect them. But the chances that they're within a thousand or ten thousand years of our level are very, very small – that would be enormously coincidental.

'You can work out some numbers, but you quickly convince yourself that if you pick up a signal, it's from a society that's thousands of millions of years beyond our own.

'It could be that they're so far advanced that we're not capable of recognizing their presence. It's like when I walk into the backyard and the ants look up but don't really look up and I'm looking down at them – and they're completely unaware of that.'

Shostak suspects that the reason we haven't detected alien life thus far is because we are using the wrong technology. But might there be a more profound reason? Could we be alone in the Galaxy? 'People have worried about the fact that we don't see obvious evidence of aliens, and they've suggested that this might indicate we're the smartest critters – the only intelligence – in the Galaxy.

'I would be astounded if that were true, but maybe the reason we haven't seen anything so far is that this is some sort of cosmic backwater. If someone were to deposit me in the middle of Nevada, I might look around – and think, 'I don't see any phone wires, I don't see any cars, I don't see any houses'. I might conclude that, I don't know where I am, but this continent is clearly uninhabited. Perhaps the Earth is located in a sort of a Nevada of the Milky Way, and that's why we haven't seen anything so far.'

With the discovery of planets around other stars, the search for extraterrestrial intelligence has ramped up into a much higher gear. And the driving force behind the quest is to obtain a better understanding of ourselves. 'I think the most important thing we're learning here in our search for planets around other stars is that we're learning about our home,' observes Geoff Marcy. 'Is our Solar System unique, or is it some common run-of-the-mill type of planetary system? Is our Earth a precious little planet with liquid water on the surface, or is it a dime a dozen? We don't really know.'

'I mean, stepping back from all the details of these discoveries, the bottom line is very simple,' adds Seth Shostak. 'When I was a kid, people talked about the existence of other planets as if they were a possibility and probably a bit of a long shot. And now planets are all over the place. They're getting longer and longer lists – it's raining planets, as the guys who do this like to say.

'What it all means is that it's another step down the yellow brick road on the way to Oz. For us, Oz is the discovery that we're not the only thinking beings in the Universe.'

Wayward

chapter four

asteroids may end up on a collision course with Earth

Worlds

They called themselves the Celestial Police. Headed up by a Hungarian baron and a German chief magistrate, the self-imposed duty of this eighteenth-century constabulary was to patrol the dark alleyways of the Solar System, on the trail of a missing planet.

Little did they know it at the time, but this was the beginning of an investigation that would lead, two centuries on, to the search for a culprit that is not just a 'missing person' on the cosmic scene, but has the potential to become a mass-murderer – of the entire human race.

From the beginning, it was not going to be an easy case. The evidence in hand was purely circumstantial. And the self-styled celestial detectives were pursuing an enquiry that flew in the face of conventional wisdom. After all, in those days everybody knew of six big, bright planets circling the Sun, from Mercury out to Saturn. If there was an extra planet, why couldn't we see it?

The Celestial Police based their case on statistics. The Director of the Berlin Observatory, Johannes Bode, had been investigating how the planets are distributed in the Solar System. He published a formula that predicted how far the successive planets lay from the Sun. Except the formula broke down in the middle. It worked for Mercury, Venus, Earth and Mars; and for Jupiter and Saturn. But it also predicted a planet between Mars and Jupiter. And there was nothing there, except a yawning void.

Until 1781, most astronomers dismissed Bode's flawed 'law' of planetary distances as a fluke. But in that year, an astronomer in England discovered a planet lying beyond Saturn – so distant that it was only on the verge of visibility. The discoverer, William Herschel, wanted to call the new planet George's Star, in honour of the British King, but this piece of blatant flattery was thwarted when Bode suggested an alternative that caught on internationally – Uranus, the Greek god of the sky.

More importantly, Bode realized the new planet would fit his 'law' exactly, if he extended it beyond Saturn. Suddenly, the gap between Mars and Jupiter became not just an oddity, but the suspected lair of another planet too faint to be seen.

In Hungary, Baron Franz Xaver von Zach was inspired to search for the new world, using a small telescope. More than a decade later, he had seen thousands of stars, but no new planet. So, in the autumn of 1800, he recruited five other astronomers in northern Europe, to form the Celestial Police force.

Each member was to examine a particular part of the sky, draw up a star-chart and then repeatedly patrol his own patch 'to confirm the unchanging state of his district, or the presence of each wandering foreign guest. Through such a strictly organized policing of the heavens, divided into twenty-four sections, we hoped eventually to track down this planet, which had so long escaped our scrutiny.'

To ensure a thorough combing of the heavens, some twenty-four astronomers were required. So the original team sent invitations to astronomers throughout Europe, asking them to join the patrol. One headed towards Sicily, where the Palermo

Observatory – the most southerly in Europe – was perfectly positioned to search the southern regions of the sky. It was also headed up by a remarkable astronomer, who was the proud possessor of the most outstanding telescope of the age.

Sicily was a backwater in Europe, so poor and underprivileged that the government had recently started a determined campaign to improve its schools and colleges. The King of the Two Sicilies, based in Naples, personally backed the idea of building an astronomical observatory. But none of the famous names in European astronomy would consider moving to Palermo.

In desperation, the authorities invited an itinerant Italian maths lecturer to take charge. When he arrived in Sicily, Giuseppe Piazzi knew nothing of the heavens; yet in a few years, he had become one of Europe's leading astronomers.

Piazzi realized he had one edge over the established observatories of northern Europe. From Palermo, he could see further into the southern skies, as well as enjoying the privilege of viewing the northern stars. In fact, he was in a position to chart more stars than anyone else in the Old World. And, for his great project of mapping the sky, Piazzi was determined to have the best telescope in the world.

His quest took him to London, to the workshop of a celebrated Yorkshire-born instrument maker. Jesse Ramsden was world-famous for his precision theodolites, sextants and telescopes. The Ordnance Survey's first mapping of the British Isles relied on his exquisite theodolites. But he was equally famous for his slow delivery. A telescope destined for the Dublin Observatory, for instance, eventually turned up twenty-seven years late.

The determined Italian was having none of that. On top of ordering a telescope more precise than even Ramsden had made before, Piazzi haunted the Yorkshireman's workshop, hounding him every day until the instrument was complete.

Palermo's unique southern location would have put Piazzi high up on von Zach's recruitment list anyway; his renowned telescope was an ideal weapon for the celestial police in their hunt for the missing planet. Indeed, while von Zach's letter was still on its way, Piazzi unintentionally fingered the culprit.

ABOVE *Urania, the muse of astronomy, indicates to Giuseppe Piazzi that there's a new planet awaiting his attention.*

Two hundred years on, Palermo still basks in the glory of that moment – a turning point in the history of astronomy. David Hughes is an English astronomer visiting Palermo for the bicentenary celebrations, and the Ramsden telescope is almost an object of pilgrimage for him. 'This is a superb example of English engineering of the 1780s; it's a most beautiful telescope, and a telescope which it's worth travelling distances to look at, just for its own beauty. And it's an instrument of great accuracy.'

The telescope's elegant brass tube is pivoted in the centre of two vertical brass circles, five feet across. Like a curved ruler, scales around the edge show precisely where the telescope is pointing.

'What really impresses me about this instrument,' Hughes enthuses, 'is its state of preservation – it's essentially as it was then. Look – you can actually see the streaks of varnish that were put there over two hundred years ago. You can imagine the joy of Giuseppe Piazzi at receiving this instrument; there was no astronomer in Europe that didn't want a telescope like this, but he got it here in Palermo.'

With his new acquisition from Ramsden's workshop, Piazzi immediately began his celestial Ordnance Survey, mapping precisely where the stars lie in the sky. 'Cataloguing stellar positions was then the routine of the vast majority of astronomers,' Hughes explains. 'They weren't astrophysicists; they weren't interested in the compositions of the stars, they were just interested in where they were and how they were moving.'

The sky was clear over Palermo on the fateful night of 1 January 1801. Piazzi's self-imposed schedule of celestial cartography had brought him to the stars in the constellation of Taurus, the bull.

'He realized on this wonderful night,' says Hughes, 'that he could see a star that wasn't expected to be there. And so, like any normal astronomer, he thought, "Tomorrow I'll check, see if it still exists." And so on 2 January he trained his telescope

ABOVE *The most accurate astronomical instrument of its time, the slim tube of the Palermo telescope is dwarfed by its massive brass mounting.*

again on this position south of the Pleiades in Taurus, and there was this new star – but it had moved slightly.' Most moving objects in the sky are comets, and that was Piazzi's first conclusion. But night after night, he followed its motion, and it didn't move like a comet would. 'From the speed it moved, he realized it was in a much more circular orbit, and from the speed with which it moved against the background stars he realized again that this object was between Mars and Jupiter. He had found what these other people were looking for – the missing object between Mars and Jupiter.'

Even before he was recruited into the Celestial Police, Piazzi had made a citizen's arrest. But could he retain the culprit? The new object was heading into the glare of the Sun, and wouldn't reappear for several months. Piazzi measured its position over and over, to try and predict its future path. Even though Piazzi had been a professor of mathematics, the formulae of his time weren't up to the task: the culprit was about to make a getaway.

Fortunately, the Celestial Police were also in touch with a young German mathematical genius, Karl Friedrich Gauss. He had just invented a new way of analysing measurements in science. It was an unbeatable alliance: the precision measurements from Piazzi's telescope, and Gauss's new theory. The culprit's future path could be predicted with confidence. And as the new celestial object emerged from the Sun's glare, it was apprehended by the head of the heavenly cops, Baron von Zach.

With its orbit now securely known, astronomers could add this new world to the inventory of the Solar System. As its discoverer, Piazzi had the right to choose a name. He chose Ceres Ferdinandea – Ceres for the patron goddess of Sicily, and Ferdinandea to flatter Ferdinand IV, the King of the Two Sicilies. He had learned nothing from William Herschel's unsuccessful attempt to call his new planet George's Star just twenty years earlier; the regal reference was again quickly dropped by the world's astronomers.

But the new world, Ceres, was certainly not forgotten. This oddity was clearly smaller than any known planet, only a few hundred miles across. Indeed, in a telescope it looked so small that it appeared as merely a speck of light, no different from a star. Astronomers began to refer to it either as a 'minor planet' or an 'asteroid' (star-like object).

The Celestial Police decided not to disband immediately. After all, who knew what other shady characters might be lurking out there? And their optimism was well rewarded. Over the next few years they tracked down three more asteroids orbiting between Mars and Jupiter. The Solar System would never look the same again.

'The view of the Solar System changed dramatically, starting late in the Age of Enlightenment,' says Steve Ostro, an asteroid expert at NASA. 'Up till then, remember, there were just the planets that were known at that time, and selected moons of those planets.'

As well as the discovery of the asteroids, astronomers had realized that comets were 'out there' in the Solar System, not mere meteorological phenomena as the ancient Greeks had believed.

'Meanwhile there was a growing interest in reports that had accumulated over history, of rocks falling from the sky,' Ostro continues. 'Some of these reports were apocryphal; some were hoaxes; and some of them were accurate, it turns out in retrospect.' Finally, in the nineteenth century a shower of stones fell in broad daylight on a village in France. With hundreds of witnesses and so many cosmic stones, scientists were forced to acknowledge that space is riddled with rocks that can fall to Earth. NASA's Steve Ostro admits ruefully: 'The last stronghold – where people were loath to admit the reality of meteorites – was allegedly in the United States.

'So within the course of one long generation, or at most two generations, we went from thinking of the Solar System as the Sun and six planets with assorted moons, to a Solar System that contains comets – whose characteristics were still unknown – and asteroids and also some little objects that occasionally fall to Earth.

'As time has progressed, space seemingly has got fuller and fuller, and we realize that these minor bodies are moving around the whole Solar System rather like a swarm of bees.'

And they can pack a deadly sting. In the twentieth century, astronomers acknowledged that it wasn't only small stones that can fall to Earth. Some of the asteroids are wont to stray from their homeland, beyond the orbit of Mars, and can approach the Earth rather too close for comfort.

'In the early part of the twentieth century,' says Ostro, 'the first Earth-crossing asteroids were discovered. And in the 1960s, a close approach of Icarus, an Earth-crossing asteroid, sparked some interest – both popular and scientific – in the possibility that an asteroid could collide with the Earth.'

In the 1980s, American geologist Gene Shoemaker pointed out that our planet bears the scars of many ancient celestial collisions, and raised everyone's awareness that it could happen again. 'Shoemaker first estimated the enormous numbers of asteroids in the Earth-crossing population,' Ostro remarks, 'and that's when people became aware of the impact hazard as a process, external to the Earth, that could dramatically affect civilization and even destroy it.'

Over the years, the census of small bodies – both remote and dangerously close – has increased rapidly as astronomers have used finer and finer nets to scoop them out of the darkness of space. From watching the sky through a telescope astronomers moved on to taking celestial photographs. On a long exposure of the night sky, the stars appear as points of light, while the moving asteroids are smeared out into a streak of light. But not everyone was happy when a new asteroid swam into their view.

'Many astronomers were trying to take photographs of stars, wanting all these

wonderful star images on their photographic plate,' Hughes explains. 'What they didn't want were these streaks – in essence, they ruined the plate, and the astronomers had to go back and do it again. So asteroids were referred to as "the vermin of space". '

Today, astronomers track down the vermin with a finer net still, using large telescopes, light-sensitive electronic chips and sophisticated computers. 'The catalogues today contain well over twenty thousand asteroids with known orbits,' Hughes estimates, 'but we've still got a great deal further to go. I calculate that when it comes to asteroids bigger than one kilometre in diameter, there are about a thousand million of them between Mars and Jupiter. So those we've discovered so far are just peanuts.'

The proliferation of asteroids has enormously complicated the name game. At first, astronomers followed Piazzi's example and named asteroids after classical

ABOVE *Vermin of the sky! Asteroid Toutatis appears as a streak of light in this long-exposure view of the stars.*

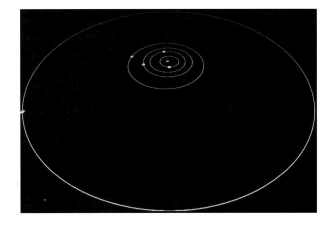

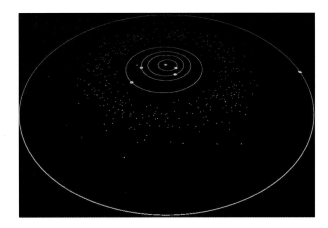

goddesses: Ceres, Pallas, Juno, Vesta; some gods and heroes also crept in. But there was a limit to even the population of Mount Olympus. As numbers grew into the hundreds and then the thousands, asteroid discoverers started to honour great humans – Einstein, Beethoven, Gagarin – and countries and cities, including Chicago, Russia and Uganda.

But in the late twentieth century, some of the names started to raise eyebrows. One turned out to be named after an astronomer's mistress. Mr Spock appeared in the heavens: not the *Star Trek* character, but the discoverer's cat. More policing was required, and the International Astronomical Union set up a committee to judge whether a suggested recipient was really worthy of the honour. One of the authors of this book recently survived the selection process, and has been elected to the heavens as asteroid Heather; fortunately, this object currently poses no threat to the Earth.

Over the past twenty years, though, astronomers have tracked down some three hundred asteroids that do stray into the realm of the planets, and may end up on a collision course with our planet. 'We know the basic statistics of the Earth-crossing asteroids,' says Steve Ostro, 'and the numbers are truly enormous. There are a thousand with a size of a kilometre or more; and perhaps a hundred million that are bigger than ten metres.'

Astronomers are now on a quest to understand the enemy. But it's far from easy. Even out in the asteroid belt, where most of the minor planets reside, space is

LEFT (TOP TO BOTTOM) *The growing number of asteroids known in 1801 (top), 1901 (middle) and 2001 (bottom), with the orbits of the first five planets out to Jupiter.*

so big that these worlds are thinly spread. And they are so small that from one asteroid, you'd be hard pushed to make out any of the others. 'If you were fortunate enough to go on a spacecraft through the asteroid belt,' says David Hughes, 'without careful planning you would see absolutely nothing.'

When NASA has sent probes to the distant giant planets, like Jupiter and Saturn, they have passed safely through the asteroid belt without hitting even a sliver of rock, let alone an asteroid. So forget the film-makers' favourite scene of space-pilots ducking and weaving through a maelstrom of giant hurtling rocks! 'In fact, the spacecraft that have seen asteroids as they've been from Earth to Jupiter have had their orbits very, very carefully planned,' continues Hughes. 'In particular, the *Galileo* mission to Jupiter went past two wonderful asteroids, Gaspra and Ida, specifically on purpose and took wonderful images.'

The first close-up pictures of the asteroidal threat were snapped in the early 1990s. The images from *Galileo* confirmed what astronomers had long suspected: these asteroids were shaped like cosmic potatoes, tumbling end over end. They are too small for their gravity to smooth out the bumps and dips into a spherical ball.

Only the biggest asteroids – over two hundred miles across – have enough gravity to pull themselves into a globe. NASA astronomers have turned the Hubble Space Telescope towards these monsters in the minor planet league, but even its powerful eye can discern only vaguely the shape of these tiny worlds.

Just down the road from the Hubble's headquarters in Baltimore, Maryland, Bob Farquhar was getting frustrated by the space telescope's tantalizing hints and the

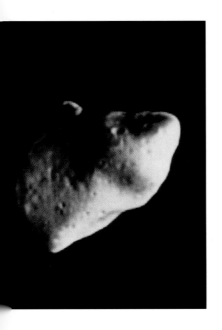 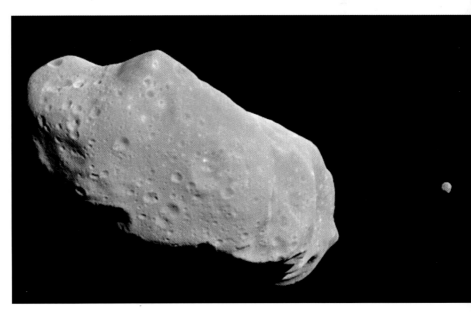

ABOVE LEFT *The first close-up picture of an asteroid was this portrait of Gaspra, imaged by the speeding* Galileo *spacecraft.*

ABOVE RIGHT Ida *was the second asteroid to sit for* Galileo's *camera: the spacecraft revealed this minor planet has a tiny moon, Dactyl* (right).

fleeting glimpses from space probes zooming through the asteroid belt. To really understand these minor planets, he wanted to send a space probe that would catch up with an asteroid and investigate a potentially lethal world at close quarters.

Now at Baltimore's Johns Hopkins University, Farquhar had previously worked for NASA, and had built up a reputation for daring and unusual space missions. When Halley's Comet came past in the mid-1980s, NASA had built no spacecraft to investigate the renowned celestial visitor. So Farquhar instructed a satellite already in interplanetary space to fly past the Moon, where it could pick up enough speed from the Moon's gravity to catapult it out to this once-in-a-lifetime target.

'The thing I'm interested in more than anything else is doing things for the first time,' Farquhar admits. 'I don't like to work on things that have been done before, and I haven't – I've been lucky.'

It wasn't luck so much as painstaking planning and innovative ideas that led Farquhar to become the first person outside NASA itself to direct a NASA-funded mission. His team at Johns Hopkins University put in a bid under a new initiative for 'faster, better, cheaper' spacecraft. Among the other winners – all from within NASA – were the famous Mars Pathfinder mission that took a charismatic unmanned rover to the surface of the Red Planet, and Lunar Prospector, which found evidence for ice on the Moon.

ABOVE *From the distance of the Earth, the large asteroid Vesta appears only as a misshapen blur – even in this view from the eagle-eyed Hubble Space Telescope.*

Farquhar's robot craft *NEAR* – the Near-Earth Asteroid Rendezvous – was headed for a potato-shaped lump of rock called Eros. After the death of the world's leading expert on asteroid impacts, Gene Shoemaker, in a freak car accident, the spacecraft was rechristened *NEAR-Shoemaker*. Gene Shoemaker had once said that he wanted to tap a geologist's hammer on Eros, and Farquhar would try to realize that wish, in the closest way he could.

While the previous asteroidal mugshots had involved 'vermin' way out in the asteroid belt, between Mars and Jupiter, Eros is a comparatively near neighbour. Its orbit crosses the path of Mars, and almost reaches the Earth. In a million years' time – a short time in the history of life on our planet – Eros's shrinking orbit will start crossing the path of the Earth. In the far future, the god of love may become our nemesis.

For Farquhar, Eros's dangerous proximity had a silver lining; in theory, it should be easier to reach than most other asteroids. But he couldn't launch *NEAR-Shoemaker* straight to its destination: a direct path would have sent the spacecraft whizzing past Eros. Farquhar needed to manoeuvre the spacecraft into a path that would bring it up to Eros gradually. *NEAR-Shoemaker* could then slip into orbit around the asteroid, held by Eros's tiny gravitational tug. As a result, he planned a looping path for the spacecraft, heading out beyond Mars, then back past the Earth before its rendezvous with Eros.

'A year before launch,' recalls this wizard of interplanetary billiards, 'I identified an opportunity to fly past another asteroid on our way to Eros.' As *NEAR-Shoemaker* spun out beyond Mars on its first loop, it could pay a fleeting call on a large asteroid called Mathilde. While the asteroids imaged by the Jupiter-bound probe *Galileo* had been made of shiny rock, the light reflected from Mathilde suggested it was as dark as coal, and probably laden with tar and sooty substances. 'We weren't supposed to do that,' smiles Farquhar. 'Almost everybody was against it, but I talked them around.' At this far point in its interplanetary loop, *NEAR-Shoemaker* was designed to be hibernating, not operating its camera. Farquhar's colleagues were concerned the spacecraft wouldn't be able to cope; after a complex and risky manoeuvre, it would point its camera the wrong way and return a picture of empty space.

'It was probably the most difficult fly-by of all time,' admits Farquhar. 'And it cost us some extra fuel – about ten extra kilograms. The head of our science team, Andy Cheng, got to me about it: he said, "Oh, we're wasting a lot of fuel chasing this other asteroid." ' As it turned out, Farquhar's plan paid off. 'It was exciting, because I really wondered if we'd even see the thing. And when that first picture came in, that was unbelievable: you could see the little craters and everything. I have to admit I even began to cry a bit, you know, it was emotional.' And the initially sceptical Andy Cheng admits it piqued his scientific curiosity: 'Mathilde has an extremely low density, suggesting that it's been battered into a rubble pile.'

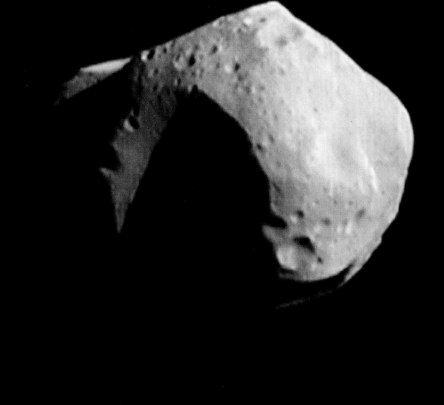

After looping round the Earth, *NEAR-Shoemaker* was set to meet Eros. All that was needed was a blast from its onboard rocket engine. Farquhar well remembers Black Sunday, 20 December 1998. 'You know, nine times out of ten, the motor's going to work. Well, it didn't. We botched the rendezvous – that was a real screw-up.

'The funny thing was that the Japanese had a spacecraft going to Mars,' Farquhar continues, 'with an engine that needed to fire to get them out of Earth orbit. And the same day ours malfunctioned, theirs did too – it was a bad day for bipropellant engines! So they had a problem, but they didn't have any contingency plans built in. Those guys were up day and night for about two weeks trying to figure what to do. We were very leisurely about it; we had it all worked out; we knew exactly what we were going to do.'

His meticulous planning had allowed even for the rocket failure: Farquhar had sent the spacecraft skimming through Earth's gravity in such a way that there would be another chance to encounter Eros a year later. In the meantime, *NEAR-Shoemaker* sailed past Eros, without a chance of entering orbit on this loop. Ever resourceful, Farquhar snapped some shots as the spacecraft sped past.

ABOVE *Mathilde was revealed to be a lump of cosmic charcoal – a carbon-rich asteroid full of holes – by the* NEAR-Shoemaker *spaceprobe.*

OPPOSITE *Asteroid Eros, seen in stunning close-up by* NEAR-Shoemaker, *is a cosmic rock pitted with tiny craters and strewn with house-sized boulders.*

A year on, as *NEAR-Shoemaker* headed for its eventual rendezvous, Farquhar discovered an unexpected coincidence. The trajectory would bring the spacecraft to Eros in mid-February 2000. By tweaking its path slightly, he arranged that the celestial rendezvous would occur on 14 February. It was a wonderful public-relations opportunity: 'a Valentine's Day's date with the asteroid named after the Greek god of love.'

As *NEAR-Shoemaker* slipped into Eros's gravitational embrace, it was another big first for Bob Farquhar. 'But I wasn't as emotional as at the Mathilde fly-by,' he admits. 'Everybody was hyping it up – you know, "We're going into orbit around this thing," like it's a big risk. And I thought, "Come on, there's nothing to it." We had it all planned out, so it didn't seem to me to be a big deal.'

For Andy Cheng it was the beginning of a year-long scientific bonanza. The spacecraft snapped Eros from all angles, sending 160,000 pictures back home. These images revealed 100,000 craters pockmarking the tiny world, and over one million boulders bigger than a house.

The pictures showed Eros wasn't a rubble-pile like Mathilde. 'It's a more or less intact rock that has perhaps been broken off some bigger object in the past,' Cheng concludes. Subsequent collisions have blown out the smaller craters in Eros, without breaking it apart, and scattered boulders over its surface.

And observations from *NEAR-Shoemaker* have allowed Cheng to probe the history of Eros. He has discovered that it has changed very little since the birth of the Solar System, over four billion years ago. 'The materials you find in Eros have not been completely melted in that time – these rocks are older than Earth.'

Even so, these were remote soundings. Bob Farquhar had one final ambition: to get up close and personal with Eros. 'Early on, I'd been talking about a crash-landing on the surface,' he says. 'But every time I would even hint about it, people would shut me up. But I thought, "Why argue with them? I'm going to be in charge of this thing at the end, anyway, so I'll do it, I'll just do it."'

When the time came, his plan had grown even more audacious. Not just to crash onto Eros, but to make a controlled landing, softly and safely. It would be the ultimate asteroid first.

'After we had achieved all the primary goals of the mission,' he argues, 'then it's crazy not to take the risk and land. I didn't want to do a stunt; I wanted to get something for my money.' His scientific pay-off would come from the spacecraft's gamma-ray spectrometer; nestled into the asteroid's surface, it could check out the material making up Eros in unprecedented detail.

Bob Farquhar gave a live press conference as *NEAR-Shoemaker* began its glide path. In the next room – unknown to him – was NASA's Administrator, Dan Goldin. NASA was then smarting from a series of spacecraft failures, including a Mars mission which was lost because the controllers had confused metres and feet. Goldin was panicking that *NEAR-Shoemaker* would not survive the landing, and would be branded another NASA failure.

Ever optimistic, Farquhar refused to utter the weasel words 'controlled crash'; his view-graph clearly proclaimed the aim was a soft landing. Goldin's face betrayed his anger as he came into the conference room – to hear Farquhar announce that *NEAR-Shoemaker* had landed safely on Eros. The conference video reveals the NASA Administrator's abrupt transformation from anger to relief, to congratulation, as he shook Farquhar's hand and patted him on the back.

In its final brush with Eros, *NEAR-Shoemaker* touched down at a mere four miles per hour – the gentlest descent in the history of space travel – on an extreme world. Exposed to radiation and micro-meteorites, the rocks on Eros's surface are 'space-weathered' to dust. And at only twenty-one miles long and eight miles wide, Eros has the weakest gravity of any world where a spacecraft has landed. Tipping the scales at half a tonne on Earth, *NEAR-Shoemaker* weighed less than a pound on Eros:

OPPOSITE *This sixty-ton ingot of cosmic iron, near Hoba West in Namibia, is the largest meteorite ever discovered on Earth.*

a high-jumper could leap a mile high on Eros, and risk putting herself into orbit!

Farquhar's first aim in landing on Eros was to get more and more detailed pictures from the descending spacecraft – and this he achieved, in spades. 'Our goal was to get forty pictures,' he says, 'and we ended up with sixty-nine. And after we landed, when we started getting data back, we couldn't believe it – it was great.' From its final resting place, *NEAR-Shoemaker* was sending back its ultimate trophy: the detailed analysis from its gamma-ray spectrometer.

Eros has opened a door into the mysterious world of asteroids, and their links with the most primitive rocks in the Solar System. 'But Eros is only one asteroid,' emphasizes Andy Cheng, 'and there are many, many other asteroids – and many types of asteroid. I'm convinced there's still a lot more we need to learn about them. Ideally, we'd like to bring samples directly back to Earth – it wouldn't even be terribly expensive to do this.'

In the meantime, nature has already arranged free delivery of selected asteroid samples: those once-controversial rocks that fall from space. David Hughes sums up the connection. 'To me asteroids are not only fascinating in their own right, but they're also associated with meteorites. So not only can we stand on Earth and look at asteroids out there, we also have plenty of examples of bits of asteroids coming to visit us here on Earth.'

Most of these splinters from the asteroid belt are too small to do us much harm: there's no record of anyone being killed by a meteorite, though an American woman ended with a badly bruised hip when a cosmic rock fell through the roof of her house in the 1950s. But – until Cheng realizes his dream of prospecting a large trawl of asteroids – meteorites provide a vital clue to the nature of the enemy that is poised to strike us from space.

Some are solid rock, chipped off from asteroids like Eros. Superficially, they look like any old rock you might pick up on Earth. Under the microscope, though, the cosmic stones contain tiny specks that stand out to a geologist's trained eye. 'These tiny specks – calcium, aluminium inclusions – may be the remnants of star explosions,' explains Steve Ostro. 'They are Rosetta Stones for understanding the origin of the Solar System.'

ABOVE *Frozen into a lump of ice is a small fragment of the meteorite that fell on Yukon's Tagish Lake in January 2000.*

Another type of meteorite is made of pure metal. These broken shards provide chilling evidence of cosmic violence in the asteroid belt. The iron meteorites were forged in the hellish interior of minor planets – a miniature version of the Earth's white-hot metallic core. A mighty cataclysm must have destroyed these asteroids, in cosmic collisions so violent they smashed whole worlds into fragments, spilling their contents into space.

A close encounter with a meteorite of a third kind rudely awakened the inhabitants of a remote region of the Yukon, in western Canada, on 18 January 2000. At 8.43 a.m., local resident Jim Brook was out walking when something brilliant caught his eye. 'It started over above those trees,' he recalls, 'and it took several seconds to move right across the sky. Then there was this massive explosion. I could feel myself shake with it.'

Unknown to Brook, a lump of cosmic debris had been on collision course with the Earth. Hitting the atmosphere, it broke up and rained fragments of meteorite down on the ground. A week later, Jim Brook was the first to find some remains of the giant meteorite. As he drove his truck home across frozen Tagish Lake, 'I came round the corner and found these lumps of rock – smelling of sulphur.'

But diligent searches have revealed only a few pounds of the original meteorite. Don Brownlee from the University of Washington, at Seattle, explains why: 'It was a hundred-and-fifty-tonne object, but actually pretty fragile: only ten kilograms were found on the ground. The whole thing just blew up – it was brighter than the Sun – and it just wasn't strong enough to survive as big pieces. I mean, this Tagish Lake meteorite was mostly busted up into dust.'

The fragments of cosmic projectile that survived at Tagish Lake are neither solid rock nor iron. They are crumbly, dark and rich in carbon: sooty stones from space. It's a description that also fits the first asteroid encountered by the *NEAR-Shoemaker* spacecraft, as Andy Cheng confirms: 'We think that Mathilde is related to meteorites called carbonaceous chondrites, like the Tagish Lake fall.'

The asteroid–meteorite connection provides the forensic clues that have allowed astronomers to understand the origin of the potential killers that stalk our local region of space. They are a deadly legacy from the birth of the Solar System.

'Four and a half billion years ago,' David Hughes explains, 'the Sun was surrounded by a flat disc, containing dust and water vapour. The dust condensed to form rocky lumps, and out of this you made Mercury, Venus, Earth, Mars – and you're potentially trying to form another planet where the asteroid belt is.'

But a growing monster was lurking in the background. In a region five times as far from the Sun as the Earth lies, the Solar System was cold enough for the water vapour to freeze into innumerable snowflakes that swirled through space. And these snowflakes quickly built up – in their zillions – into the giant of the Solar System, the massive planet Jupiter.

'Jupiter affected the formation of this planet that was trying to grow in the region of the asteroid belt,' Hughes continues. 'Instead of this material coming together to form an individual planet, Jupiter's gravity perturbed them so they started smashing each other up. That's why – between Mars and Jupiter – we have literally a collection of bits.'

These bits are the ammunition in the cosmic shooting range that still endangers the Earth. But we live at a comparatively halcyon time. In the early days of the Solar System, Jupiter flung most of the asteroidal matter away in a deadly machine-gun spray. 'A thousand times more material was thrown out of that region than actually ended up in the asteroid belt,' estimates Brownlee, 'and Jupiter totally destroyed the planet that was trying to form there.'

Mars, the next planet in, was barely better at resisting Jupiter's gravitational turmoil. 'Mars has only ten per cent the mass of the Earth,' Brownlee continues, 'so Jupiter's gravity presumably threw out something like ninety per cent of the material from this region also.'

The scars from this reign of terror are easy to find. 'These particles being flung around by Jupiter's gravity went all over the Solar System and hit things,' says Hughes. 'Wherever you look – on the Moon, on Mars, on the satellites of Jupiter and Saturn – you see craters produced by asteroid bits that started their life between Mars and Jupiter.'

Much as the worlds of the inner Solar System have suffered from this bombardment, many more lethal fragments were thrown outwards. Beyond the orbits of all the planets, they survive to the present day, as a second cosmic arsenal poised to wreak havoc on the worlds of the Solar System.

'There's a huge swarm of objects a long way from the Sun, literally getting on for halfway between our Sun and nearby stars,' Hughes explains. 'These objects are really dirty snowballs – they're made of snow and dirt.'

Each dirty snowball has the potential to flower into a beautiful awesome comet: astronomers call them cometary nuclei. But how does a cometary nucleus differ from an asteroid? Opinions are divided.

Hughes: 'Asteroids and comets are completely different things. Asteroids are lumps of rock and metal, produced by large planet-sized objects smashing up. Comets are dirty snowballs, simply building blocks that were on the way towards forming the cores of the giant planets Jupiter, Saturn, Uranus and Neptune.'

Brownlee: 'I think it's fascinating that people always talk about comets and asteroids as if they're completely different things – and I personally don't. In fact, they may have been remarkably similar, early on in the history of the Solar System. If you'd looked at the asteroid belt at that time, most of those asteroids would have had "comet tails". They were warm enough to melt water on the inside, so they must have been outgassing like gangbusters.'

What's undisputed is that present-day cometary nuclei – lying so far from the Sun – are currently stored in deep-freeze. But occasionally the gravity of a passing star will dislodge a comet nucleus from its perch. It starts to fall in towards the Sun, speeding towards the vulnerable planets near the centre of the Solar System.

'Imagine I was sitting on one of these cometary nuclei,' says Hughes, 'then as it moves in towards the Sun I would be moving faster and faster. And the snow inside is getting heated up by the Sun's energy. There's one rather exciting transition, just as I get closer to the Sun than Jupiter; the snow is warm enough to convert to gas, and so the surface of the comet suddenly starts emitting gas. When I reach the closest point to the Sun, the surface is disappearing as the snow melts before my eyes.'

The gas – mainly steam – billows out into space as a vast glowing cloud. The dirt in the 'dirty snowball' is carried with these gas jets as palls of dark dust. The Sun's energy pushes the steam and dust away into space, in long tails that may stretch over a million miles. 'In those classic pictures of comets,' enthuses Don Brownlee, 'there are two tails – a dust tail and a gas tail. The rocky material is going off in one direction; and the blue tail, that's the water leaving the comet.'

A fully fledged comet, with two brilliant tails, is a rare sight in the sky. But millions of people were treated to this once-in-a-lifetime experience late in the 1990s. The story began with an informal band of latter-day celestial police, constantly patrolling the heavens for intruders from beyond the known planets and checking the behaviour of those already known.

One member of that international volunteer force is Alan Hale. His professional work involves investigating stars way beyond the Solar System, and his fascination with comets is a spare-time activity. For many years he had combined a search for new comets with a regular check on comets already in the catalogues.

'On the night of 22/23 July 1995,' he recalls, 'I had planned to observe two comets. I finished with the first one just before midnight, and had to wait about an hour and half before the second one rose high enough to look at.' To pass the time, he turned his telescope towards the constellation of Sagittarius, the archer. Here the sky is rich with star clusters – swarms of tightly packed stars. One of Hale's favourites is the star cluster M70.

'When I turned to M70 I saw a fuzzy object in the same field, and almost immediately suspected a comet.' He contacted the International Astronomical Union, who later confirmed that his fuzzy object was indeed a stranger from the depths of the Solar System. 'I love the irony,' he grins. 'I've spent over four hundred hours of my life looking for comets, and haven't found anything, and now – suddenly – when I'm not looking for one, I get one dumped on my lap!'

Even more ironically, the celestial visitor was also discovered – the same night – by an amateur astronomer who had never even hunted for comets. Tom Bopp is a shift supervisor

in a construction materials company in Phoenix, Arizona, but his passion is the heavens – and, in particular, faint and fuzzy patches like galaxies, nebulae and star clusters.

'On that night,' he recounts, 'some friends and I headed out into the dark deserts west of Stanfield, Arizona.' Around 11 p.m., his friend Jim Stevens suggested checking out some of the star clusters in Sagittarius. 'I was watching M70 slowly drift across the field of view,' Bopp continues, 'when a slight glow appeared at the eastern edge. I repositioned the scope to centre on the object, and called over to Jim.'

Jim Stevens checked on all the charts they had to hand, but nothing was marked – no galaxy, no nebula, no star cluster. What else would look fuzzy through Bopp's telescope? The group knew the answer all too well. 'The moment Jim said, "We might have something," excitement began to grow among our group. And I breathed a silent prayer thanking God for his wondrous Creation.'

Back home, Bopp had a more mundane communication to make – to the International Astronomical Union. 'When they telephoned and said, "Congratulations, Tom, I believe you discovered a new comet," that was one of the most exciting moments of my life.'

By tradition, comet discoverers are more privileged than astronomers who track down asteroids. Instead of merely having the right to name the new object, each comet discoverer is personally commemorated in the appellation given to the new celestial visitor: in this case, Comet Hale–Bopp.

Over the following year, Hale–Bopp swung in towards the Sun, growing ever bigger and brighter. 'Nobody could miss it,' David Hughes enthuses. 'I can remember walking out into my garden; the one obvious thing in the sky was this wonderful comet, and the stars were as nothing.'

If a twenty-first century astronomer can be so emotionally moved, what were our ancestors to make of the awe-inspiring sight of a brilliant comet? In cultures around the world, comets have had a deep symbolism. When a comet appeared over China in AD 712, the emperor Jui-tsung chose to abdicate because he thought it would otherwise bring disaster to the country. And William Shakespeare summed up the Western view in *Julius Caesar*: 'When beggars die, there are no comets seen; the heavens themselves blaze forth the death of princes.'

These days, princes can breathe comparatively safely when a comet comes by – at a reasonable distance, like Hale–Bopp. But the whole of humanity should quail if a comet is found to be heading directly our way. The huge comet Hale–Bopp gave us a two-year warning of its approach; with a small comet mankind might have only a few months to prepare for Armageddon.

It's a menace that will be with us for the foreseeable future, no matter how powerful telescopes and computers develop to warn us of the peril from space. Though equally dangerous to us, comets and asteroids are in different leagues when it comes

OPPOSITE *Comet Hale-Bopp was the sky-sight of 1997. It sported two tails, a yellow trail of dust and a blue stream of water vapour.*

to predicting their chance of impacting our planet. In the next few decades, astronomers hope to chart all the dangerous asteroids flying around our part of the Solar System. But there's no hope of tracking all the distant cometary nuclei that could one day pose a threat to the Earth.

We can only rely on the few months' grace a comet gives us by sprouting its tell-tale glowing head and tail. Minor planets give us no such warning. 'Asteroids can whistle past us more or less unnoticed,' Hughes comments, 'while a comet nucleus of the same size will surround itself with a huge atmosphere and tail – no one can miss it.'

When that threat comes, our only hope will be to intercept the icy harbinger of doom – and attempt to destroy it. But how do we annihilate a virtually unknown object? We must, once again, get to know the enemy. That chance came in 1986.

'Halley's Comet comes back every seventy-six years,' explains Hughes, 'and that year it was due to pass the Sun. In 1986, the Space Age was well developed. And here was this famous comet coming back to the inner Solar System, and we had rockets and space technology, we didn't just have to stand on Earth and look at it; we could actually go there.'

An international armada was prepared, in a five-way race to Halley's Comet. Ironically, the world's premier space agency wasn't even on the starting blocks. NASA had prepared a comet mission so complex and expensive that it was cancelled even before it was built. That was when the ever-ingenious Bob Farquhar used the Moon's gravity to fling an old spacecraft through the tail of another comet – the first comet-encounter in space history – and then on to Halley's Comet. Even Farquhar couldn't arrange snapshots of the famous visitor, though, as the *International Comet Explorer* carried no camera.

Photography was left to the Japanese, the Russians and the European Space Agency. They all launched novel missions. One of the Japanese spacecraft took pictures of Halley's Comet in ultraviolet light, impossible from the ground. Two Russian probes sped past Venus, dropping balloons into the planet's storm-tossed clouds, before heading into the heart of Halley's Comet. Their TV cameras sent back the first pictures of a solid nucleus at the heart of the vast glowing cloud; pictures that were, however, blurred and indistinct.

It was the European *Giotto* craft that stole the show. David Hughes takes up the story. 'This mission was named after the famous Italian artist who painted Halley's Comet in 1301 as the Star of Bethlehem. It passed Halley's Comet at a velocity of a hundred and fifty thousand miles an hour!'

In that instant, in the early morning of 13 March 1986, *Giotto* beamed back to Earth its uniquely intimate portrait of Halley's Comet. On TV screens around the world, viewers were puzzled by the real-time views, in lurid false colour. But the scientists were ecstatic: when decoded, the *Giotto* image was sharp, clear – and thrilling.

OPPOSITE *Steam erupts from the 'dirty snowball' at the heart of Halley's Comet, in this unique close-up shot from the plucky* Giotto *spaceprobe.*

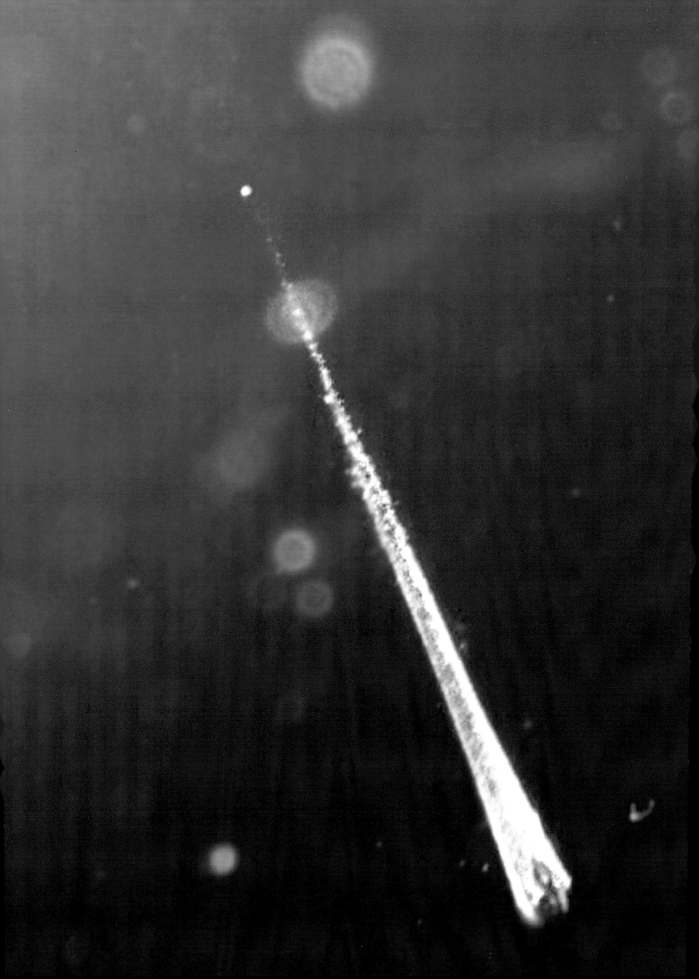

'It was a wonderful picture,' Hughes raves, 'showing great jets of gas activity. We could work out its shape; we could see hills and valleys and depressions on this avocado-pear-shaped cometary nucleus. We could try and work out the way it was spinning; we could work out how much mass it was losing.'

Most of all, scientists were surprised by its colour: black – as black as coal, as dark as sable velvet. In this 'dirty snowball', it turned out, the snow is all on the inside, with the dirt on the outside. Halley's Comet was little more than a giant cosmic choc-ice.

For Hughes, even a single picture of a cometary nucleus is priceless. 'Just imagine knowing that human beings lived on Earth and you having no pictures of them. And then someone comes along and gives you just one picture of a human being – what an exciting thing to look at!

'Now of course you want to look at more; you want to know if all human beings are the same, you want to know how they evolve and change. Now we're stuck with just one picture of one comet, and we need a few more shots in our picture-books of comets. We think we know quite a bit about comets, but in essence all we do know is one comet in reasonable detail.'

Scientists may hunger for more information on comets for their own sakes, but future knowledge will also be essential when – one day – we need to safeguard our planet from the curse of a comet on collision course.

And several space missions are now setting out to check out a wider array of comets. Already on its way is NASA's *Stardust* mission. Skimming past Comet Wild-2 in January 2004, six times faster than a rifle bullet, it will scoop up a sample of the comet's dust and bring it back to waiting scientists on Earth.

Meanwhile, scientists across the Atlantic are keen to subject a comet to the kind of intimate scrutiny that the *NEAR-Shoemaker* spacecraft performed on asteroid Eros. The European mission *Rosetta* is set for a rendezvous with Comet Wirtanen in November 2011. *Rosetta* will orbit its nucleus as the comet speeds towards the Sun, filming the growing eruptions of steam and dust as the dirty snowball swelters in the Sun's heat.

While *NEAR-Shoemaker* landed on Eros only as an afterthought, *Rosetta* has a purpose-built lander, complete with camera for the closest views of this unknown heavenly body. 'I really do want to know what the actual physics of this comet is,' says Hughes, 'I want to know if it's full of holes, I want to know how the snow and dust has come together. And I want to know what it's like inside. It's only by landing on the surface and seeing how that surface shakes that we can start to probe into the interior.'

In the meantime, NASA has planned the mother of all shake-ups for a comet nucleus – literally a sledgehammer approach. 'The idea is very simple,' says Jay Melosh, the scientist charged with calculating the effect of NASA's comet-strike. 'We're going to take a spacecraft with three hundred and fifty kilograms of copper on it, and we're

OPPOSITE *A speeding dust particle* (upper left) *is caught safely after ploughing through a special lightweight gel, in a test of the* Stardust *mission to capture comet dust.*

going to run it into Comet Tempel-1 at ten and a half kilometres per second – and blow as big a hole as we can!'

Aptly named, the *Deep Impact* spacecraft will send back to Earth a running video of its kamikaze dive, while a sister spacecraft films the destruction. The crater is predicted to be as big as a football pitch, and seven storeys deep. Melosh is keen to point out that there's serious science involved in this cosmic raid: 'One of the mysteries of comets is what's deep inside them, and by blasting entirely through the crust and exposing the deep interior we can figure out what was inside.'

ABOVE *The* Rosetta *spaceprobe drops a landing craft on to the surface of Comet Wirtanen, in this artist's impression of the audacious 2011 rendezvous mission.*

Telescopes on Earth will tune into the gas cloud erupting from the impact. Amateur astronomers – as well as professionals – will be inviting the public to view this ultimate display of cosmic fireworks, on American Independence Day, 4 July 2005.

And some watchers may experience quiet satisfaction at the sight. After billions of years during which our planet has been at the mercy of the vermin of space, the inhabitants of planet Earth have now struck back.

Ham
th

chapter five

would have been strong enough to rupture your eardrums

mer of
e Gods

From an aeroplane window, the sunset paints ever-richer colours on the already absurdly ruddy Arizona desert as you head westwards towards Los Angeles. Racing headlong towards the setting Sun, dark shadows sweep in, making the desert moody, mysterious.

Suddenly, you spot it right below. A huge crater, a gaping wound on the desert's smooth surface – looking as fresh as if it had been blasted out yesterday. Half-filled with inky-black shadow, Arizona's Meteor Crater is a chilling reminder that our planet is vulnerable to the very real threat of bombardment from space.

Meteor Crater is a punctuation mark in the history of the Earth. And Jay Melosh is the perfect guide. Based at the Lunar and Planetary Laboratory in Tucson, at the other end of Arizona, he is an expert on interplanetary collisions.

Melosh paints a word-picture of the day that the Arizona desert experienced a hammer-blow from space. 'If you'd been around fifty thousand years ago, you'd have seen a meteorite approaching very quickly. Within a second or two it would have plunged from the top of the atmosphere into the ground – right here.'

The meteorite disappeared into the ground. And a fraction of a second later, a spray of white-hot rocks erupted high into the sky. The whole crater blossomed in the next few moments, with more rocks whistling through the air. The solid ground rocked, and a blast wave ripped through the atmosphere.

'I've been personally present at an explosion test,' confides Melosh, 'where you could feel the ground shock and the air shock. The air shock is actually a double blast, as you get from a supersonic plane. And you can see it coming. When the explosion occurs, you see what looks like an expanding shimmering bubble, looking like a bubble of window glass.'

The 'bubble' is compressed air from the blast, squeezed so that it has a different refractive index from ordinary air. 'And it's expanding tremendously fast,' Melosh continues. 'Everything is completely silent until it gets to you; and then you hear this tremendous double boom of the shock wave.'

If you'd been in the Arizona desert that fateful day, though, even a distance of several miles would not have provided much safety. The meteorite erupted with the force of twenty megatons of TNT – over a thousand Hiroshima atom bombs.

'The blast would have been strong enough to rupture your eardrums,' Melosh continues, 'the heat from the incandescent ejecta would have set you on fire. Your eyes would be burned out; your hair would be burned off; and you probably wouldn't feel anything, very shortly after that.'

The excavation of Meteor Crater, in fact, wasn't at all like the hole you get by dropping a stone into thick mud. The meteorite burrows its way underground; as it's stopped, its enormous energy of motion is converted into explosive power. It's travelling so fast that a ton of speeding meteorite provides more explosive energy than a ton of TNT.

OPPOSITE *Half-filled with shadow – enhancing its fresh, stark contours – Arizona's Meteor Crater is a salutary reminder that we are still at the mercy of being struck by objects from space.*

And the hole it's blasted out is an awesome sight, from the air or from the ground. Approach it from the legendary Route 66 – now Interstate 40 – as you head west towards Flagstaff, and you first see what looks like a low ridge rising from the desert. Not till you reach the top of the ridge do you see the gaping hole that lies beyond; three-quarters of a mile across, and five hundred feet deep. Its owners proudly sign it as 'The Planet's most penetrating Natural Attraction'.

Jay Melosh tosses pebbles into the depths of Meteor Crater as he puts that claim in perspective. 'In some ways it's not a very remarkable crater – it's not very big – yet it's exquisitely preserved and easy to visit. And it was the first crater to be recognized as being formed by impact; to be so obviously of impact origin that it convinced even the most sceptical scientists that a meteorite had made it.'

In the 1880s, sheep farmers living in this – then remote – part of Arizona reported finding nuggets of silver. They turned out to be pieces of iron, mixed with nickel to make an alloy so tough that it couldn't be broken open with standard chisels. It was identical to the material that makes up iron meteorites.

Still, the claim of a meteoritic crater was controversial to begin with. The landscape of northern Arizona fuelled an argument between opposing geologists. Near Meteor Crater lie the volcanoes of the San Francisco peaks, and smaller craters that are clearly the work of exploding volcanoes. Some geologists argued that this crater, too, was volcanic; by coincidence, a scattering of small iron meteorites had happened to fall to Earth in its vicinity.

ABOVE *Astronomer Jay Melosh on the rim of Meteor Crater. The crater is three-quarters of a mile across, and is marketed as 'The Planet's most penetrating Natural Attraction'.*

The argument was settled unequivocally in the 1950s, by the first geologist to span the gulf between the Earth and the other planets. Gene Shoemaker was prospecting for uranium in Colorado and Utah when he came across the strange crater amongst Arizona's volcanic landscapes. He was entranced. From then on, craters and celestial impacts were to be the core of his scientific life.

'One of the things you see in the crater rim,' says Melosh, 'is that the rock units have been turned upside-down. So what used to be the surface inside the crater is now lying upside-down, essentially its back, and the sequence of rocks has been inverted. That was first recognized by Gene Shoemaker in the 1950s.'

The craters from nuclear explosions also have over-turned rims, as Shoemaker pointed out, but it's not something you find with volcanic craters. It requires an intense pressure from a small underground explosion: a buried nuclear warhead, or a meteorite that's burrowed into the ground before erupting. And, crucially, Shoemaker also discovered rare minerals in the crater, which can only be made by explosions far more powerful than a volcanic eruption.

Gene Shoemaker applied to NASA to become the first geologist on a world that's a crater-lover's dream come true: the pockmarked surface of the Moon. To his lifelong regret, he was rejected from the astronaut corps because of a minor health problem. To make up for his grounding, he personally taught the Apollo astronauts their geology, including training sessions at Meteor Crater.

And this scar on Earth's surface was ringing warning bells in Shoemaker's ears, long before anyone else became concerned. If our planet had been hit before, it would be hit again: there were plenty more balls in the game of interplanetary billiards that the Earth must play in its journey around the Sun. The projectile that had blasted out Meteor Crater was the size of a large office block. Big and scary when it's falling on your head; but a minnow on the scale of interplanetary space. Even if there had been astronomers about at the time, no one would have seen it coming.

So Gene Shoemaker and his wife Carolyn set up their own search for wayward bodies in the Solar System, using a small telescope at south California's Palomar Mountain. In the early 1990s, they were joined by David Levy, an amateur astronomer from Tucson. During the years that she scoured the photographs taken by Levy and her husband, Carolyn Shoemaker discovered more comets than anyone else in history.

One night in 1993, the team turned up a truly remarkable object: 'a squashed comet' was how it first struck Carolyn. Astronomers quickly realized that this comet – the ninth discovery by the team – had had the misfortune to come under the influence of the giant planet Jupiter.

'The story of Comet Shoemaker–Levy 9 is absolutely fascinating,' says leading British comet expert David Hughes. 'In essence what they'd discovered was twenty-one bits that had been produced by the break-up of a comet because that comet had got too close to Jupiter. It was pulled apart by Jupiter's gravity.'

The mighty planet tossed the pieces of comet out again, like a cat playing with a mouse. And Jupiter wasn't about to let its prey go. Looping round in space, Comet Shoemaker–Levy 9 was destined to smash into the giant planet a year later. Astronomers around the world had their telescopes ready to observe Jupiter when the time came. Because the fragments of comet were spread out, the celestial fireworks would be spread out over six days.

'It was a wonderful time in astronomy,' Hughes enthuses. 'We'd been saying that comets and asteroids have the potential of hitting planets; and now we could stand back and see them do just that. You saw material being ejected from Jupiter, with great infrared flashes of energy. And this really hammered home to Earth-watchers the danger of cometary collisions. I mean, if bits can hit Jupiter, they can jolly well hit Earth as well!'

Hughes's ambivalent attitude to impacts reflects their role in the history of the Solar System. Cosmic collisions are a fact of life in our part of the Universe – wayward comets and asteroids have been impacting our planet since its birth. But they are not all doom and gloom.

'Basically, every atom on Earth has been through a hyper-velocity impact,' says Melosh. 'Everything that composes you and me has been through the centre of an impact of some kind or another.'

'In the broadest sense we're alive *because* of them,' elaborates Steve Ostro, a leading NASA asteroid researcher. 'Comets and asteroids can be the destroyers; but they also played a role in our creation. You could construct a mythology in which they are simultaneously the destroyer and the creator.'

This two-faced cosmic power began its give-and-take relationship with the infant planet Earth almost as soon as the planet began to take shape.

'When the Earth first formed,' explains impact expert David Kring, 'it was alone, we didn't have a Moon. But one of the largest impact events ever to affect the Earth involved an object about twice the size of the Moon. When that collision occurred between the two planets, it launched a huge plume of rock and vapour into space that went into orbit around the Earth. And within a very short space of time, that ejecta accreted into our Moon.'

The Earth, reeling from the shock, was a red-hot ball of molten rock. Exposed to the chill of space, its planet-wide oceans of lava began to cool, and the surface congealed into a skin of solid rock – like custard that's been left too long. It was a harsh, desiccated world: all our planet's original air and water had been blasted into space.

But the even-handed deity of cosmic impacts was set to redress the balance. A gentler hail of smaller asteroids and comets now began to deliver water to our dry

OPPOSITE *Asteroids and comets can even wreak vengeance on mighty Jupiter. This image reveals the glowing wounds inflicted by Comet Shoemaker-Levy 9.*

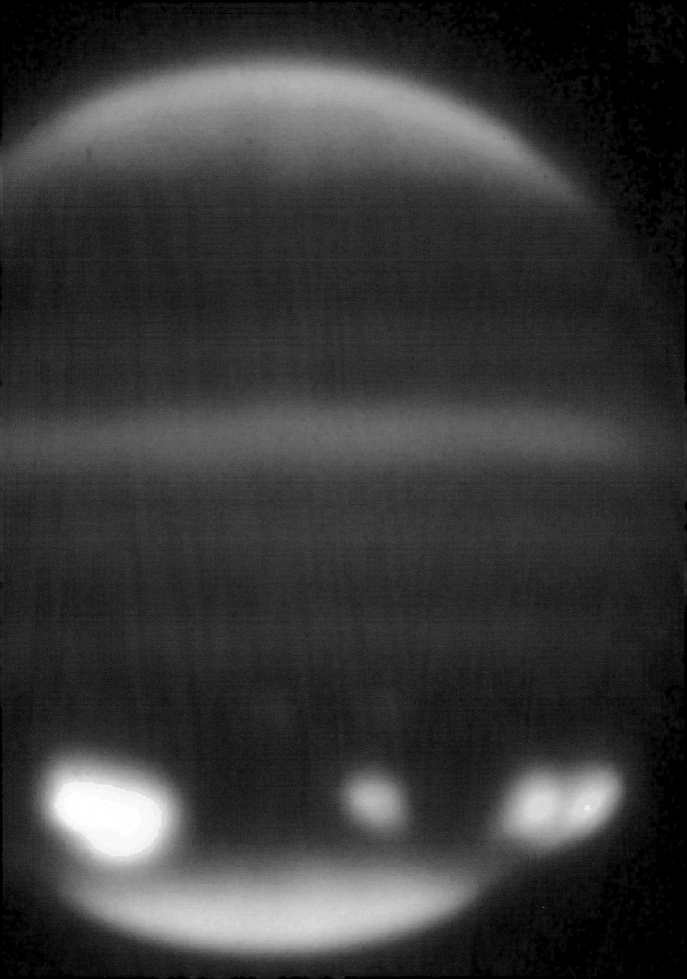

planet. Don Brownlee of the University of Washington in Seattle has been investigating.

'We know that there's forty thousand tonnes of material falling on the Earth every year,' he calculates. 'That's the rate today – early in the Earth's history the amount of comet and asteroid-like materials were actually much higher than now.'

Comets are lumps of frozen water, mixed with specks of rock. And most asteroids – though they look like solid rock – also have water sequestered within their crystals. 'Comets may be half water, asteroids ten to twenty per cent water,' says Brownlee. 'And big impactors penetrate right into the mantle of the Earth. So they can inject water into the mantle rocks, in addition to steam into the atmosphere.'

The result was a flood of astronomical proportions – lasting not just forty days and forty nights, but for millions of years on end. From a bone-dry planet, the Earth became a water world.

'Yes, the Earth could have been a totally marine planet, without any land,' asserts Brownlee. And that has implications for other 'earths' out there in the Cosmos: 'My guess is that's the most typical kind of terrestrial planet, a water-covered planet.'

Fortunately for us humans and our kin who live on dry land, the Earth recovered from its early drowning. From the two extremes – a hot desiccated world and a totally marine planet – the Earth, Goldilocks-like, ended up neither too dry nor too wet. It has enough water to fill the oceans' basins, but sufficiently little that dry continents can rise above sea level. What swung the balance, once again, were cosmic impacts.

On a clear moonlit night, you can see the evidence for this third onslaught for yourself. During this final major act of violence in the birth of the planets, half a dozen comets or asteroids sculpted the 'face' of the Man in the Moon.

ABOVE *Ultimate impact: billions of years ago, a body the size of Mars collided with Earth* (top left). *Out of the swirling debris* (centre) *was born our Moon* (bottom right).

His distorted features have intrigued people for as long as humans have walked on Earth. According to one early account, the Man in the Moon was a Sabbath-breaker, stoned to death by Moses for collecting sticks on the holy day of the week. Later, he makes an appearance in Shakespeare's *The Tempest*, in the scene where Caliban is astonished to discover visitors on his secluded island. In reply to Caliban's wondering enquiry, 'Hast thou not dropped from heaven?' Stephano teases him: 'Out of the moon, I do assure thee: I was the man i' the moon when time was.'

In the seventeenth century, astronomers turned the first telescopes to the Moon, and found a more prosaic answer. The 'eyes', 'nose' and 'mouth' are large dark plains set in the Moon's mountainous landscape. Confusingly, their names are based on the Latin word *mare*, meaning 'sea': we now know that the lunar 'seas' are among the driest plains in the Solar System.

Neil Armstrong's 'giant leap for mankind', in 1969, put his feet squarely into the Man in the Moon's left eyeball – Mare Tranquillitatis, the Sea of Tranquillity. It was the first of six expeditions to bring home a prized trophy of lunar rocks – a third of a tonne in all, much of it still in NASA vaults waiting to be analysed. Led by Gene Shoemaker, NASA's geologists were convinced that the Moon's rocks would be survivors from the very earliest days of the Solar System. Like the oldest meteorites, they were confidently expected to date back some 4,600 million years.

'We found that the Moon contains many old rocks,' reveals Don Brownlee, 'but no rocks that go back to its earliest history.' Most of the lunar rocks are a 'mere' 3,900 million years old. At this time, long after its birth, mighty hammer-blows from space reconstructed the surface of the Moon. They gouged out half a dozen vast craters; filling up with dark lava, they became the main features of the Man in the Moon.

'We call this the period of heavy bombardment,' Brownlee continues. 'The source of this heavy bombardment could have been comets being cleaned out by perturbations by Neptune and Uranus, or it could have been something more simple – a single body that came into the inner Solar System and broke up.'

ABOVE *Comets like Ikeya-Seki have a dual personality. As well as causing destruction by their impacts, they also bring water. In its infancy, Earth changed from a hot, bone-dry planet into a world covered with oceans.*

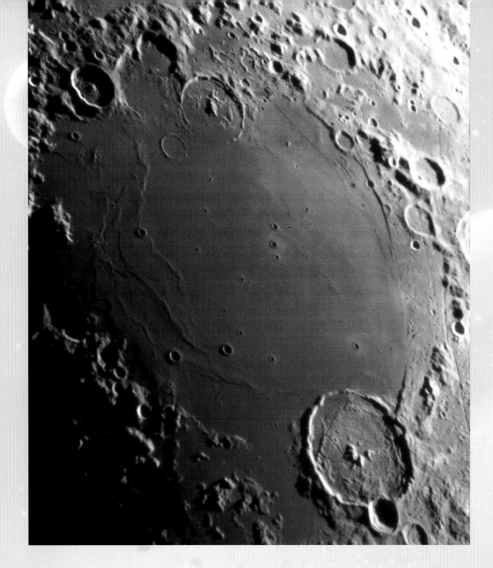

Either way, the Moon was battered for a period of seven hundred million years, from its birth in the Big Splash through to the final frenzy of the period of heavy bombardment. Though geologic processes on Earth have all but erased this period from the record on our planet, there's no doubt the watery Earth too was exposed to this onslaught.

'These giant impacts can also blast away oceans,' explains Brownlee. 'During this period – and especially during the heavy bombardment – these impacts eroded our oceans. So the ocean we have left may be only a small residual to the total amount of water we had at the beginning.'

The first seven hundred million years was also a bad time for any life that was trying to form on the Earth, as each impact blasted whole oceans into space, heating our planet's surface to boiling point. 'This has been called the period of impact frustration of life, with the idea that there were individual impacts that were capable of sterilizing the planet. So if life evolved on Earth, it would have to re-evolve after each of these events.'

ABOVE *The crater Gassendi flanks the Moon's two hundred-and-forty-mile-wide Mare Humorum. The Moon's lava-filled 'seas' – which make up the face of the 'Man in the Moon' – are the legacy of the last period of super-impacts, 3.8 billion years ago.*

It's probably no coincidence, then, that oldest fossils of life date back to a period shortly after the heavy bombardment. These microbes – which chanced to form when the heavy bombardment ceased – were the first that could survive and multiply without the threat of imminent destruction.

But where did that life come from? Cosmic impacts again played a Janus role. On the one hand, the giant hammer-blows repeatedly destroyed life on the young Earth; on the other, they reseeded our planet with the raw materials of life.

Steve Ostro elaborates. 'Small bodies – whether they were comets or asteroids is immaterial – delivered the volatiles and the organics, which led to the origin of life and the chemistry of nucleic acids and long chains of amino acids.'

In the 1950s, chemists tried to make these basic building blocks of life from the gases in the Earth's original atmosphere. And the experiment was an astounding success. Amino acids – the basis of proteins – and the nucleic acids that build into DNA were churned out by the bucket-load when experimenters discharged artificial lightning through a mix of gases that simulated the Earth's early atmosphere.

'So we have two sources,' says Brownlee. 'They could be made locally on Earth; or they could be brought in.'

But the made-on-Earth theory has recently been suffering serious setbacks. Scientists have revised their ideas on our planet's original air. Instead of being rich in methane and ammonia, they now think it was made mainly of carbon dioxide. And, when you strike a spark through that mixture, the gases refuse to metamorphose into any interesting organic molecules.

'So there are times when people worry about how much of this organic material could have been produced on Earth,' sums up Brownlee, 'but the amount that comes in from space is real – there's nothing you can do to prevent the arrival of these organics. Organic material arrives from space right now, and the arrival rate was presumably thousands of times higher early in the history of the Solar System.'

The rate may be lower now, but it is still enough to have given the residents of the small town of Murchison, north of Melbourne in Australia, a rude awakening in September 1969. A large meteorite thundered through the morning sky, its explosion unleashing a shrapnel-burst of rocks from space. More than two hundred pounds were collected, and rushed to Houston, Texas. Here, NASA scientists had just completed the world's most sophisticated laboratory for space rocks, designed to handle the precious cargo that the Apollo astronauts were returning from the Moon.

And this free sample from space turned out to be far more interesting than all of the dry sterile lunar rocks put together. The Murchison meteorite is one-eighth water; and it is laden with organic molecules. Among them are traces of sixty-two different kinds of amino acids. This tally includes almost all the varieties that build up the proteins in our bodies.

'There's nothing you can do to prevent the arrival of those organic precursors from space,' Brownlee emphasizes. 'All this organic material was dumped on the Earth and it may have played a key role.'

Not just a key role, but the star role, if Brownlee's latest research is anything to go by. He's found that tiny particles of cosmic dust get fried as they enter Earth's atmosphere, and shed carbon in the form of microscopic cornflakes. 'So there's a continuous rain of black carbon falling on the Earth. This black micro-porous carbon is the world's best sponge for organic material, and it also has little iron bits in it that are catalytically active. So it could have played an incredibly important role in the origin of life.

'Or,' he adds philosophically, 'it could have played no role whatsoever. And how would you ever know? 'Cause no one kept any records! We have these two extremes; so it's fun; it's a little bit frustrating, but it's more intriguing than frustrating.'

The question of the origin of life has certainly frustrated the world's most eminent scientific minds for over a century and a half. Charles Darwin characteristically opened up the whole debate by suggesting that life first arose in a 'warm little pond'. Now the field has been thrown wide open. Not warm ponds, say some scientists, but

ABOVE *Swathes of dark cosmic dust cut through the glowing light from a brilliant nebula. This material constitutes life's raw materials – and it was almost certainly brought to Earth by impacts.*

superheated jets of water escaping from volcanic vents deep on the ocean floor. Others invoke the vast cold reaches of Earth's early oceans; cracks deep in the Earth; or volcanic hot springs.

David Kring, from the Lunar and Planetary Laboratory in Tucson, Arizona, looks to the heavens for life's genesis. And not just for the manna of organic material raining down from space. Kring suggests that cosmic impacts kick-started life on Earth.

'If we look at the genetic make-up of life today,' he says, 'and try to find out what our common ancestors were like, we realize that those organisms lived and thrived in hot-water environments. And of course an impact event deposits a lot of energy into the crust of the planet, heating it up and causing fluids in the Earth's crust to circulate.'

Today, such hydrothermal systems are driven by the heat of volcanic rocks hidden way below the surface – in Yellowstone National Park in Wyoming, for instance, or Rotorua in New Zealand. But in our planet's early history, missiles from space would have been just as effective. And the impact on Earth would have smashed open cracks in the Earth's crust, where the hot elixir of life could flow.

It's more than coincidence, in Kring's view, that the first signs of life appeared soon after the curtain came down on the great cosmic bombardment of 3,900 million years ago.

'This extraordinary flurry of impact events completely resurfaced the Earth – and of course our neighbouring Moon,' he says. 'It deposited a tremendous amount of energy in the Earth's crust, which caused water and other fluids to circulate. And those are exactly the types of system where pre-biotic chemistry can evolve to form life. And in fact it's the type of environment where the most primitive forms of life on Earth would have actually thrived.'

To check out his ideas, Kring has come to Sudbury crater, a vast ancient impact crater in Canada. He reaches out for a rock that's crossed by a fine pattern of light-coloured lines. 'These veins were created when fluids flowed through the rock, and from these fluid systems minerals precipitated on the walls, producing this lovely lacy texture. From these rocks we can see how the fluids migrated through the rocks, and we can calculate the lifetimes of these hydrothermal systems.

'While the impact event was very brief, lasting only minutes, these hydrothermal systems may have survived for a hundred thousand to perhaps a million years, giving ample opportunity for life to evolve.'

Sudbury crater itself came too late to be a crucible of life on Earth. It is a 'mere' 1,800 million years old, and by then microbial life had long been disporting itself in the oceans. But it's a good analogy for the earliest craters, which have long since been erased by the Earth's ceaseless geology: eroded away, filled in or buried deep beneath later sediments.

In fact, even Sudbury is scarcely recognizable as a crater today. The long flat pastoral valley, surrounded by rolling forested hills, has none of the arresting splendour of Arizona's Meteor Crater. Part of the problem is sheer scale: even the whittled-away remnant of this giant scar stretches for thirty miles. To travel from one geologically exciting outcrop within the crater to another, Kring must drive for an hour.

In the late nineteenth century, prospectors were drawn to Sudbury by its ores of iron and copper. Then, in 1886, a chemist discovered that these ores contained vast amounts of nickel. In fact, Sudbury held far more nickel than the world could use at that time.

One of the mine owners had, coincidentally, come across metal meteorites, and was aware that their iron–nickel alloy was too tough to mark with either a file or cold chisel. He persuaded the US government that nickel-steel would make virtually invulnerable armour plating – and Sudbury's fortune was made.

The mine owners may have been pleased, but the geologists were an unhappy bunch. They couldn't work out the area's unique geology. The nickel-rich ores formed a giant oval ring, in the hills surrounding Sudbury's flat central plains. And there was no sign of volcanic activity – usually the agent for concentrating rare metals into rich veins of ore.

Then, in 1964, American geologist Robert Dietz dropped a bombshell. Even before Gene Shoemaker came on the scene, Dietz had been one of the first scientists to argue that the craters of the Moon were blasted out by impacts from space – and that the lunar 'seas' were lava plains flooding even bigger craters. While searching for the equivalent of a *mare* on Earth, he came to Sudbury. There he found some strange-shaped rocks that were the fingerprint of an explosion far greater than any force on Earth can create.

Drawing up by a rocky outcrop beside the road, David Kring points out the giveaway formation. 'That cone-shaped object in the rock is what we call a shatter-cone. This one's less than a metre across, but some can be the size of a full-grown

ABOVE *Calling-card from space: this shatter-cone formation of pulverized rocks at Sudbury, Canada, was created by the shock wave from the collision of an asteroid with the ground.*

person. What's happened is that the shock wave from the impact has literally shattered the existing rocks here – transformed them, destroyed some of their crystal structure.'

From these clues, Kring can recreate that fateful day when Canada was subjected to a far mightier hammer-blow than the impact that created Arizona's Meteor Crater. It began when an asteroid twice the size of Mount Everest was headed to an inexorable encounter with planet Earth.

'Here at Sudbury, an asteroid ten to twenty kilometres across slammed into the Earth,' he begins. 'It flashed through the atmosphere as an extraordinary sky-splitting blinding fireball. It penetrated the surface of the Earth to a depth equal to its own diameter. The energy of that event was tremendous, and it was released in a huge explosion that destroyed the impacting asteroid.'

ABOVE *Artist's impression of the formation of the Sudbury meteor crater. A superheated blast-wave erupts into the atmosphere as the huge asteroid is totally destroyed by the force of the impact.*

The energy of that impact was almost impossible to imagine – the equivalent of a mind-numbing one hundred trillion tonnes of TNT. That's six million times more explosive than the Mount St Helen's volcanic eruption in 1980; and six million times larger than the United States' largest nuclear weapons blast.

'The energy also destroyed a large portion of the Earth's crust here,' Kring continues. 'It was shattered, melted and vaporized – the impact caused the rock to behave more like a fluid, like toothpaste. We ended up with a shallow crater two hundred kilometres across.' That's as big as the craters you can see on the Moon with binoculars. And it wasn't just a hole punched in the existing rock.

'New rock was created here,' explains Kring. 'Within just a few minutes a huge sequence of the Earth's crust was melted, forming a layer of impact melt two and a half kilometres deep.' It was quickly topped with a layer of broken fragments a kilometre deep, as rocks blown out from the explosion rained back to Earth. Kring calculates that some of these fragments travelled almost as far as the Moon.

From a wooded hillside, Kring contrasts the placid scene today with that hellish moment almost two billion years ago. 'We're sitting on the south rim of what survives of the Sudbury impact structure,' he says, 'and it really is an extraordinary site. We're actually seeing the guts of an impact crater.

'The part of the impact crater we're sitting in used to be many kilometres below the surface of the Earth,' he continues. 'The top of the crater was well up in the sky above us – but erosional processes have scraped the top of the crater away. Think of the crater as a coffee-cup, and basically the top three-quarters of the cup have been removed.'

For Sudbury's miners, the bottom line is that the erosion has laid bare the rich mineral seams, distilled in the deep cauldron of molten rocks. For geologist David Kring, it's a unique chance to check out the underbelly of an impact crater: 'In the case of a younger, less-eroded, crater we wouldn't be able to see this – it would be impossible for us to drill this deep.'

And for Luann Becker, from the University of Hawaii, Sudbury provides a stepping stone on her way to understanding the greatest mystery of life on Earth: the sudden mass extinctions that came close to wiping out all living creatures on our planet.

'It turns out that over the history of evolution on our planet,' she explains, 'there have been some very severe extinction events. You're talking about two-thirds to almost everything on the Earth going extinct, in a very short period of time geologically speaking. We're talking about tens of thousands of years; maybe a few hundred thousand years – compared to billions of years, that's nothing.'

What's rare, sudden and brings unparalleled disaster to the Earth? Cosmic impacts, say many scientists. But to prove their point, they need to put together a case as convincing as the prosecution in a legal trial.

'What we're doing is becoming part bloodhound, part detective,' explains Becker. 'We're going into a geologic record that's dominated by terrestrial rocks, and we're trying to pluck out that proverbial needle in a haystack – something that was left behind by an impact event.'

Over the past twenty years, Becker and her colleagues have assembled a more or less watertight case that an impact was indeed responsible for the most famous massacre of all – the death of the dinosaurs.

The first solid evidence came from the picturesque medieval town of Gubbio, in central Italy. A scenic gorge, complete with an ancient aqueduct, has cut through layers of rock that were laid down during that critical period when the dinosaurs made their dramatic exit, accompanied by two-thirds of the other species on Earth.

'There's a very distinct boundary, a change in the actual rock type at that point in geologic time,' says Becker. 'And it's marked by a knife-thin layer of clay.' In 1979, American physicist Luis Alvarez took some of this unusual Italian rock back home to

ABOVE *Death of the dinosaurs: fragments of a six-mile diameter asteroid rain into the atmosphere above the oblivious creatures. Had they not been made extinct, you wouldn't be here to read this book today.*

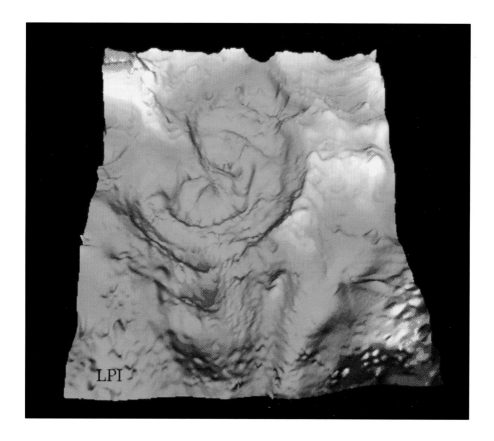

the University of Berkeley in California, and checked what it was made of. To his immense surprise, the thin layer of clay was rich in a very unusual element, iridium.

'The most compelling part of the story,' Becker argues, 'is that iridium is not very common in rocks on the Earth. But it is associated with asteroids and meteorites – that's the needle that Alvarez drew out of the haystack.'

David Kring has been investigating that same layer of geologic history in the rocks of Colorado. Here, the odd rocks marking the boundary are thicker: a brown layer-cake with treacle tints. And they clinch the case for an awesome asteroid strike on the Earth.

'This rock was actually deposited sixty-five million years ago,' he explains, 'and there are basically two layers. The lower layer was formed when molten rock from the impact fell back through the atmosphere, as a fiery rain. Then there was the finer-grain solid dust, which settled out much more slowly, forming a second layer on top of it.'

What was missing, though, was the 'smoking gun' – the crater that the killer asteroid had blasted out of the Earth. As in all the best detective stories, this critical piece of evidence was unearthed in a rather unexpected way . . .

ABOVE *The huge buried crater of Chicxulub in northern Mexico – the dinosaurs' nemesis – is revealed in a precise survey of the region's gravity.*

In the late 1980s, NASA lent a hand to archaeologists who were on the trail of the springs used by the ancient Mayan civilizations of what is now Mexico. NASA satellites revealed hundreds of sink holes around the small town of Chicxulub, on the Yucatán coast. Water disappeared down the holes, to well up again at springs by the sea.

The real surprise came when the team mapped the sink holes and springs: they formed a perfect semicircle. The curve was truncated by the coastline, so most likely the water-bearing strata formed a circle, extending out under the sea. It was a distinctly un-geological-looking feature.

By coincidence, oil-prospectors were investigating the same region. Their surveys revealed a circular shape, hidden beneath more recent rocks. And their boreholes provided the vital clue. Rocks extracted from the depths revealed that this geological oddity dated from sixty-five million years ago – the death-day of the dinosaurs.

'Chicxulub – the world's most famous impact site – is now buried a kilometre beneath the Earth's surface,' says David Kring. 'When they drilled into the structure, we had rock samples that then allowed us to study the shock deformation in the rocks in our laboratories.'

'The evidence is now insurmountable,' Becker concludes, 'that this mass-extinction event was associated with a large asteroid hitting the Earth.'

On the rim of Meteor Crater, Jay Melosh describes the far more titanic disruption that was visited on the Earth at Chicxulub. 'If you were anywhere on the continent, you'd see a column of incandescent gas go up. The heat would ignite everything within a few hundred kilometres. You feel the ground shock and the air blast – and they'd destroy every higher form of life within a few thousand kilometres.'

'The local and regional effects are absolutely devastating,' concurs David Kring. 'As well as the fireball and the air blast that roars across the landscape, the impact generates tsunamis in the shallow sea. These tidal waves radiate across the Gulf of Mexico to crash onto the coast of Texas and Alabama, ramping up to heights of fifty to a hundred metres – very powerful, very large tsunamis.'

On top of the direct destruction to the Americas, the impact would indirectly affect the entire planet. 'It's these global effects that are truly important to the alteration of biologic evolution of the planet,' says Kring.

Melosh takes up the tale. 'The material that's blasted out of the crater sails up through the atmosphere. It expands out through a hole created by the expanding vapour plume and ejecta that actually makes a partial vacuum for a while. It expands high above the Earth; and then it starts falling back into the upper atmosphere.'

These specks of rock have cooled as they're shot into space. Crashing back into the atmosphere, they are heated to incandescence in a blizzard of white-hot shooting stars. 'You know you see one meteor in the sky, and it's very beautiful,' Melosh continues, 'but instead of that one brilliant meteor, imagine the whole sky full of meteors. You would feel the heat beating down from the sky, like being inside a pizza oven.'

Over the next hour, this rain of fiery death spreads out all around Earth. 'In parts of the globe it ignited wild fires,' says Kring, 'and we have evidence of the soot from those fires in the rock record today.'

Melosh goes further: 'The expected result is that everything on Earth was on fire at that time. That, I believe, is what killed off the dinosaurs. Our ancestors, the mammals, survived because they were critters like rats that lived in holes in the ground, and they could get away from this brief intense heating event. The dinosaurs couldn't find refuge in holes; and they were probably roasted in their tracks.'

And the destruction didn't stop there. Next came major climatic changes. 'We think it deposited a lot of toxic material in the Earth's atmosphere,' Kring continues, 'which may have destroyed the ozone layer and created aerosols that partially blocked sunlight. Along with dust from the impact event, these aerosols cooled the Earth's surface, before they rained out as sulphuric acid. Finally, there would have been carbon dioxide and water liberated by the impact event: over much longer timescales, these gases led to the greenhouse warming of the planet.'

The awesome destruction of Chicxulub is impossible for the human mind to comprehend. But Luann Becker believes it was not unique. 'I'm addressing the question: did this happen more than once?'

Her quest has brought her to Sudbury. The devastation here mirrors Chicxulub, though it occurred too far back in history to wipe out any higher forms of life. Becker is investigating Sudbury to discover if there's any clue in this giant impact which could be used to 'fingerprint' other suspected impacts on the Earth.

'Here we're sitting right on top of the fall-back breccia,' she enthuses, 'that's the material that was injected into the atmosphere after this rather large impacting object hit the Earth, and then plopped back down into a rather large hole it had excavated. You can see that it's dark in colour. What's really unusual for breccia is that this rock is rather carbon-rich.'

The carbon in these rocks has lured Becker on to closer familiarity, as others have been entranced by the most expensive variety of this element. 'Diamond is one pure form of carbon,' she explains, 'and graphite is another. I've been looking at a third form, which is more unusual – it has very unique properties.'

Diamond, graphite – and fullerene. Carbon of the third kind gets its name from the geodesic domes designed by the architect Buckminster Fuller in the 1960s. If you could spy on fullerene through a powerful enough microscope, you'd see that each molecule is made of sixty carbon atoms, joined to make a spherical cage. You don't have to be an architect to recognize the pattern of links between the carbon atoms: they are exactly the same as the lines on a soccer ball!

'Fullerene has been around for just over a decade now,' says Becker. 'It was discovered by a group of scientists who were blasting graphite with a laser, vaporizing it – and actually forming this closed-cage carbon structure. And I started wondering why we hadn't discovered fullerenes on the Earth.'

The laser experiments suggested that Becker needed to find somewhere on Earth where carbon had been blasted in a serious way. That search, naturally, took her to Sudbury.

'When I brought these rocks back to the lab, I ground them and drew the fullerene out with chemical solvents.' Producing a glass vial of a coloured liquid, she continues: 'We ended up with this solution, with its unique reddish wine colour. What's in here is stardust – something that's even much older than the Earth.'

Becker had made a discovery beyond her own expectations. This fullerene wasn't created on Earth as a consequence of the cosmic blast. It was an ancient hoard of stardust, hidden onboard the asteroid as it thundered towards our planet, and scattered far and wide by the impact.

'So Sudbury was the beginning of a new way to look at the extraterrestrial record. And we were very excited about the possibility of moving on in the record. So

ABOVE *Visiting the Sudbury crater in Canada, Luann Becker holds a model of fullerene – a soccer-ball of carbon atoms that links asteroid impacts to mass extinctions on Earth.*

we turned our attention to the Cretaceous–Tertiary boundary – the end of the dinosaurs – and checked out the boundary layer at some half-dozen sites around the world, from New Zealand to North America to Denmark. And we established that indeed fullerene was there right with the iridium.'

So stardust was spangled over the globe from the Sudbury impact; and also from the Chicxulub catastrophe that killed off the dinosaurs. Becker suspects that every asteroid may carry fullerenes on board. So each past asteroid calamity should leave its calling card in the rocks, as a thin layer of stardust.

'Impact events of a size that formed Sudbury or Chicxulub occur on average every hundred million years,' says David Kring. 'So, during the Phanerozoic – the period of complex life on Earth – we know we should have had about five or six impact events of that size.'

Sudbury itself is too old to count: it was blasted out before life had evolved to the stage of fish or even trilobites. Chicxulub is clearly one of the handful we'd expect. And here, the impact has clearly coincided with a mass extinction.

'In fact,' says Kring, 'during this same period we had five major mass extinction events. So one of the hypotheses we're exploring is: are these impact events correlated with extinction events every time?'

Luann Becker is now checking for any signs of a cosmic cataclysm at the time of the mother of all mass extinctions. 'The Permian–Triassic extinction event – that's about a two-hundred-and-fifty-million-year-old event in the geologic record – happens to be the most severe mass extinction in the history of life on our planet. Something like ninety per cent of all living organisms on the continents and in the oceans were extinguished – rather abruptly.'

Once again, Becker is investigating the thin layer of rocks that mark the abrupt about-turn in the history of life, and in the Earth's geological story. 'We went in, tried to get the best boundary samples we could, to isolate the fullerenes,' she says. 'And in a couple of locations – one in China, another in Japan – we think we've found evidence for fullerenes. So now we have two of the "big five" extinction events that appear to be associated with an extraterrestrial cause.'

It's an unfolding story, and Becker and her colleagues are still a long way from convincing many geologists, who prefer more traditional explanations. Some put the mass extinctions down to vast outpourings of lava; other researchers blame the rise and fall in sea-level; while yet another view invokes changes in the Earth's climate.

Steve Ostro, an astronomer who takes delight in watching asteroids whizz past the Earth, has few doubts. 'Asteroids definitely mattered in the evolution of life,' he avers. 'Periodically over the last several billion years there have been impacts that eradicated most of the life on the planet, eliminating life from ecological niches, effectively resetting the evolutionary clock, altering the playing field for the competition between various subsets of species on Earth, and irrevocably changing the direction that evolution was taking.'

'It begins to point us towards a new way of thinking about the evolution of life,' adds Becker. 'How life changes from, say, the dinosaurs to the mammals; and even before that, at the time of the Permian–Triassic when life flourished a lot in the oceans, and began to then take over the continents – and gave rise to the dinosaurs.'

'It suggests to me,' she continues, 'that we had to have a severe cleansing of life to establish a new way of life on our own planet. It may well turn out that these sort of extinction events are necessary for life to evolve to what it is today.'

David Kring elaborates. 'In the case of the Chicxulub impact event, it dramatically altered the biological evolution of the planet. If it wasn't for that event, mammals wouldn't have evolved; the human species wouldn't have evolved. So the fact that you and I are sitting here talking has a lot to do with the fact that there was this impact event in Mexico sixty-five million years ago.'

'I think it's probably likely that the dawn of human intelligence was affected in part by the motions of the sky,' muses Ostro, 'primarily the Sun and the Moon, of course, but also the wanderers identified as planets, and then the intermittent phenomena like comets, meteors, fireballs and perhaps an occasional impact.'

And with the evolution of intelligence, planet Earth at last has a species that can recognize the danger of space impacts. 'Asteroids are the most important part of our environment except for the Sun and Moon,' Ostro opines. 'In some senses they are more important, because the impact hazard is real and it will affect our future.'

'We can say, with absolute confidence,' asserts David Kring, 'that impact events like Chicxulub are going to occur again. And, more importantly, you don't need an impact that size to cause devastating global consequences. The human species is a very complex organism and probably more susceptible to alterations in the environment. You can cause global effects with an asteroid as small as a third of kilometre across.'

Each century, according to Ostro, there's a one-in-a-thousand chance that civilization will be wiped out. 'Eventually we'll find an object that is on a collision course. When that happens – and it is a matter of *when* it happens, not *if* it happens – then people will be able to go to a planetarium and read a fact-sheet. It will give the chronology of when it will collide with the Earth, how much energy will be released and what damage it could do.

'There will be uncertainties. You might have a situation where there's a fifty per cent chance it will hit in the Atlantic Ocean; a twenty per cent chance of impacting the United States; and a twenty per cent chance it'll hit Western Europe. And you might have no hope of refining those numbers for who knows how long.'

And, if the hammer of the gods is as massive as the asteroid the dinosaurs faced, then nowhere on Earth will there be a hiding place . . .

Lov

chapter six

ultimately be used as the celestial equivalent of Noah's Ark

ing the
Enemy

The moment you land in Puerto Rico, you're faced with a mass of contradictions. The island is American territory, yet its citizens can't vote in the US; and despite the fact you're in the Caribbean, the place has a Spanish, swaggering, macho feel. If you are so minded, you can buy moustache-grooming kits at the airport.

Visit the capital city, San Juan, and you'd swear you were in an ancient fortified Spanish citadel. But go just a little way down the coast road to the west and you find yourself amidst smooth, constantly watered golf resorts.

There are two, parallel coast roads on Puerto Rico's north coast, a freeway and a toll road. Most people drive on the freeway – and it shows. The average resident in Puerto Rico has two or three cars, almost without exception in the form of rusting metal boxes. The freeway is jammed with ancient Chevrolets and drivers throwing the remains of rolled-up cigarettes out of their windows.

The toll road is largely deserted, and it's a delight to drive, hugging the contours of this lush, tropical island. But the roads have one thing in common. They lead to a place that is unique on Earth: the site of the Arecibo Radio Telescope, the biggest radio telescope on the planet. One thousand feet across, it nestles into the densely wooded karst landscape in a natural limestone hollow.

It's an awesome sight. The wire-mesh metal dish is incredible enough, but the view from the gondola-shaped focus box – 515 feet above – is to die for. You get to it by going up in a cable car to an open-lattice metal trackway suspended above the dish. Grown men have trembled at the upper cable-car platform and then refused to go any further.

Puerto Ricans are deeply suspicious of Arecibo. Despite the fact that the astronomers working there have openly tried to tell all and sundry that it's a scientific research instrument, there's a feeling on the island that the dish reeks of warfare and the military.

But maybe it does deserve those connotations, in a somewhat tangential way. For the Arecibo Radio Telescope is a leading weapon in our war against the threat from space: the menace of asteroids that may collide with the Earth. NASA's Steve Ostro, who works at Arecibo, fronts the vanguard of the defenders of our planet. And like a military surveillance expert, he uses radar as his secret probe.

'In most astronomy, you go to a telescope and look at the light or the radio waves coming in. In radar, it's active astronomy: you do an experiment on the Solar System. You have a powerful transmitter, you design the signal you're going to send out, and transmit for a period of time at an asteroid. Then when the echo's starting to come back, you change to receive mode, and get the echo back.

'So you can think of it as asking a question with electromagnetic waves and getting an answer. The other way of looking at it is that you're reaching out and touching an object. We don't do passive work. We check these objects out personally

– our intelligence contacts these objects. It's like we're having a conversation.'

It's important to converse with asteroids. More important, it's essential to have an ongoing conversation. Because asteroids are always going to be a part of living on planet Earth. They represent both our past and our future. Small though they are, they have massively shaped our past history, as witnessed in the legacy of huge impacts like Arizona, Chicxulub and Sudbury. On the other hand, they have also delivered water and minerals to Earth, although not without huge cost to life. Asteroid collisions are a fact – and the threat isn't going to go away.

Many asteroids have by now been mopped up by collisions with planets, or have been flung out of the Solar System. But there are still a lot of them around – and they will undoubtedly continue to hit the Earth. So if we want to fathom what the future holds in our love–hate relationship with these mini-worlds in space, we need to get to know them intimately. We need to know what they look like, what they're made of, what they're up to. And in the far future, we might even look into ways we can use asteroids for the good of humankind.

All these thoughts are in Steve Ostro's mind when he uses the world's biggest radio telescope to hunt down the Solar System's smallest objects. When Ostro bounces radio waves off asteroids, he becomes ecstatic. 'The experience of seeing the first echo from an asteroid – getting the first images – is a stunning one emotionally,' he enthuses. 'It's magical. You see for the first time what the object is like, how big it is, and what its shape is.

ABOVE *The thousand-foot-diameter Arecibo radio telescope in Puerto Rico – the biggest in the world – leads the vanguard in the crusade to track down rogue asteroids.*

'It's almost like a *Star Trek* experience – we have the good fortune to see these things for the first time, to learn about them for the first time. It's momentous, because you know that throughout the future of humanity these objects are going to play a big role.'

This philosophical perspective leads him into other areas. 'When I'm not observing, I do the Yang-style t'ai chi, an ancient system of exercises for health and relaxation that originated in Hong Kong. In doing the form, you manipulate energy – you move energy through your body in ways that are beneficial physiologically and psychologically.

'It's a nice counterpart to what I do in my work, where we take enormous amounts of energy – a million watts is what we radiate – and move it through the Solar System tens of millions of kilometres at the speed of light. And that's how we learn about these objects which, one way or another, are going to be important to the future health of human civilization.'

Ostro's work not only reveals the shapes of asteroids: it reveals where they are, and what they're intent on doing. 'Radar's the way to go to pin these orbits down,' he explains. 'The object may be tens of millions of kilometres away, but we can measure the range to twenty metres without much difficulty. And the object may be moving towards or away from us at many kilometres a second, but we can measure the velocity to an accuracy that's like the speed of the minute hand of a clock.

'That's why radar's so powerful for refining orbits. You get to know where the object's going to be not just a few decades in the future, but where it's going to be a millennium in the future.'

The asteroids Ostro is observing stray far from the main belt that lies between Mars and Jupiter. Many of them cross the orbit of the Earth, which puts them on potential collision course with us. But there is an advantage: he can study them in close-up, and is coming up with unexpected discoveries about these miniature worlds.

Ostro has just come back from the radio telescopes at Goldstone, in the heart of California's Mojave Desert. And he's excited to follow up the observations he's made with a session at Arecibo. The object of his attentions is an asteroid with the inspiring name of 1999 KW4. 'It comes very close to Earth, a little over five million kilometres,' he explains. 'At Goldstone, we've found out enough about the object to have a more efficient strategy so as to conduct the very sensitive imaging observations at Arecibo.

'And we immediately found that this is a double object, a binary system – a small object circling a larger object, like the Earth's Moon circles it. We've looked at something like eighty-four near-Earth asteroids and we've found three binary systems in the past year. And they raise all kinds of interesting questions about where they come from and how stable they are.'

But Ostro's near-Earth asteroids are of far more than just academic interest. They pose a sinister threat to planet Earth. 'About ten years ago,' he reveals, 'we realized as

a community that the Solar System was filled with small bodies in numbers that we hadn't previously imagined. Our planet exists in a swarm of asteroids. There are maybe a thousand as large as a kilometre, over a hundred thousand as large as a hundred metres, and then perhaps a hundred million around ten metres across.

'They've been impacting our planet ever since the Solar System formed, punctuating the evolution of life. They played an active role in changing the course of evolution, altering the climate, changing ecosystems, wiping out large fractions of life on Earth, opening up ecological niches so that other lifeforms can evolve – and we're the current end product.

'That's the past. For the immediate future, they constitute a hazard that can't be ignored. It's an unusual kind of hazard – it's very low probability but very high consequence – and it's something that human civilization is going to have to deal with from here on in.

ABOVE *In California's Mojave Desert, the giant Goldstone radio dish also assists in the hunt for Earth-crossing asteroids. Note the author for scale.*

'We know enough about the collision hazards. Objects hit with enough energy to disrupt or destroy human civilization maybe once every few hundred thousand years on average. We don't know when the next event will happen. We don't even know when the last event happened, because any traces of it would have been eradicated by the normal processes on the surface of the Earth – weather, plate tectonics and so forth.'

Meteorite expert David Hughes is more upbeat about the prospects for life's survival in the aftermath of an asteroid impact. 'I'll give you a trivial example. If an asteroid hits North America, you can quite easily imagine it killing a large proportion of the population on that continent. But it's also easy to imagine people living in, say, New Zealand managing to survive. It's going to be extremely difficult – considering how many humans live on the surface of planet Earth – to wipe us out.'

Hughes also warns against unnecessary doom-mongering when it comes to predicting upcoming collisions with Earth. 'In the past few years, we've unfortunately had one or two false alarms,' he recalls, ruefully. 'One that I remember very well was an asteroid named 1997 FX11. The asteroid was observed, and the orbit was calculated. Now, as soon as you calculate an orbit you think it's accurate. You put it into a computer, and you can work out exactly where this asteroid will be next year, or in ten, twenty years' time – you name it.

'And it was discovered, much to our amazement, that in about twenty-eight years this asteroid would get so close to Earth that there was a distinct possibility there would be a collision. Now, this caused a lot of anxiety.

'What did happen, though, is that as people made more and more observations of this asteroid, the orbit improved in accuracy. And as it got more accurate, we realized that the risk of the collision had in fact gone away, and that this asteroid would zip past and miss us. When I say zip past, this means about three times further away than the Moon.'

Even so, on the cosmic scale of things, it's equivalent to a bullet from an assassin narrowly missing your ear. And although David Hughes is optimistic about the survival of the human race in the case of an impact, he absolutely acknowledges that near-Earth objects pose a major threat. 'We're seriously convinced that not only have they

ABOVE *Impacts of asteroids on Earth are the stuff of Hollywood movies. But unfortunately, they aren't fictional events – and it's only a matter of time before the next major blast.*

hit Earth, but they will hit Earth. Huge numbers of asteroids have hit us in the past, and millions of craters – literally, millions of craters – have formed on the surface.'

To defend ourselves against the threat of attack from space, we need to work out the nature of the beast that we're faced with. So what are we up against? John Lewis from the University of Arizona at Tucson describes it. 'A typical near-Earth asteroid would be about one kilometre in diameter, and we know – from radar studies and also from optical studies – that they have bizarre shapes, they're not spherical. Making a body spherical is usually a consequence of having lots of gravity, and these bodies have very little gravity.

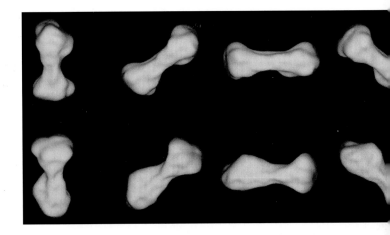

'Some of them are shaped like dumb-bells, and some of them appear to be like snowmen – stacks of roughly spherical chunks built up on top of each other like some form of sculpture. Others appear to be composed of two nearly equal-sized objects that are touching at the equator, and with a little nudge I suppose they could actually roll around each other.

'If you were to land on one, you could walk along the surface, down a crevice between them, and find yourself with an asteroid either side of you – and gravity would vanish at the point of contact between them. Furthermore, they're rotating. We found recently that many of the smaller asteroids – a hundred metres in diameter or smaller – are spinning so fast that any soil on their surfaces would be thrown away.'

'At the present moment, we can detect many of these near-Earth objects,' adds David Hughes. But he also reminds us that this doesn't diminish the magnitude of the threat they pose. 'We're in a position where we've seen Comet Shoemaker–Levy 9 strike Jupiter, and we've learned that the dinosaurs were wiped out on our planet by the impact of some asteroid or cometary object.'

But today – for the first time in Earth's history – we are not entirely powerless to protect ourselves. We can fight back against the menace from space. 'We don't have to sit back and take it,' confirms Hughes. 'With our present space development, we have the potential to do something about it. We can go there, we can try and deviate its orbit so that it misses Earth.

'Whether we go there with a large nuclear weapon and blow the asteroid to bits, or whether we go there with some other source of explosive that just pushes it to one side, that's a matter of debate. But we can do something about it nowadays, and I

ABOVE *Giant dog-bone in space: the Arecibo telescope acquired these images of the asteroid Kleopatra. It's a solid lump of nickel-iron alloy 135 miles long by fifty-eight miles across.*

think we're duty-bound to do something about it.'

And NASA is already preparing the technology that can deliver the critical warhead. The first task is to work out exactly where the killer asteroids are; and next, where the spacecraft lies. In the vastness of space, that's not an easy task. NASA is relying on Marc Rayman for its celestial navigation.

Rayman is in charge of an innovative mission called *Deep Space 1*. It's actually not designed to go anywhere in particular, but to check out some amazing and untested new ideas. 'Nobody's going to use new technologies, particularly risky ones, until someone has proved that they work,' he says. '*Deep Space 1* was designed to take the risks, so that future missions won't have to.'

Deep Space 1 is the first spacecraft to navigate its own way through space, using asteroids as its signposts. And, once Rayman has selected a target for the spacecraft, *Deep Space 1* will work out its own interception path. It's called autonomous navigation. 'The first thing you need to know is where is the body and where is it

ABOVE *The visionary NASA spaceprobe* Deep Space 1 *will use revolutionary new navigation techniques to home in on its target, plus electronic propulsion to get it there.*

going: if you're going to travel literally many billions of kilometres to get to a body that's only a dozen kilometres across, you need to know exactly where it is. In the case of Halley's Comet in 1986, the spacecraft from the former Soviet Union transmitted pictures back to Earth, which were analysed here and used to update the location of the nucleus. So the European Space Agency was able to guide *Giotto* to get it closer.'

The spacecraft is also pioneering a new form of propulsion. Instead of burning rocket fuels in an impressive fireball, *Deep Space 1*'s ion drive uses electricity to push out an eerily glowing cloud of xenon atoms. 'The first time I ever heard of this,' Rayman says, 'was in a *Star Trek* episode where aliens come and use ion propulsion to outrun the *Enterprise*!' Real rocket science has now caught up with science fiction. '*Deep Space 1* is now using ion propulsion, which is ten times more efficient than conventional chemical propulsion – so we can carry only one-tenth as much fuel. And if you combine that with autonomous navigation, you get an improvement over today's technologies which is like having your car find its own way from Los Angeles to Washington DC, arrive in a designated parking space – and do it all while getting three hundred miles per gallon!'

Deep Space 1 has all the advantages of the tortoise over the hare, as Rayman explains by comparing it to a traditional spacecraft currently in orbit around Mars. '*Mars Global Surveyor* had to fire its engine to be captured by Mars's gravity, and that took two hundred and eighty kilograms of propellant. That's a big mass to go on top of a big rocket that cost taxpayers lots of money. *Deep Space 1*, my mission, could do the same manoeuvre with only twenty kilograms of propellant. The other side is that *Mars Global Surveyor* executed this manoeuvre in twenty-two minutes, while *Deep Space 1* would take three and a half months. But ultimately it comes out ahead.'

The new technology will also buy us time to deal immediately with the threat from space. 'We don't know how to design and manufacture spacecraft quickly enough, launch them frequently enough, that we always have something ready to go. I remember Captain Kirk always giving an order to Mr Sulu or Mr Chekhov, "Launch a class-three probe, take one right off the shelf." Well, we ought to get to the point where we can do that, so when the next Hale–Bopp comes all we have to do is say, "Let's take the fourth spacecraft from the left and let's send it to this new comet." '

And one day, we'll need to come up with this kind of rapid reaction to the threat from space. David Hughes can foresee the occasion. 'If we can predict a potential impact many, many years in the future, we can then go to the asteroid, and give it a slight push, and that impact won't take place. If however the impact's going to take place not in ten years' time but say ten months' time, we then need to get there pretty quickly and push that asteroid very hard.'

He explains how we'll do it. 'The easy thing to do is simply go to the object and change its velocity. It's rather like being in a car – you put your foot on the brakes, and you change the velocity. Then you don't hit the object that's coming across your path. Or you could decide the other thing – put your foot on the accelerator and then you get past the object quicker.' But the braking or accelerating has to be done with a certain amount of force, and the sooner you see the hazard, the less force you have to apply.'

Just as there are many ways to skin a cat, there are many ways to push an asteroid. Pacing around the rim of Meteor Crater in Arizona – a salutary reminder of our vulnerability to impacts from space – Tucson-based planetary scientist Jay Melosh reviews the options. 'The most popular one you hear on the news is detonating a nuclear explosion. But what you don't hear in the popular press is that the proposal's not to land a nuclear explosive on the asteroid and blow it to smithereens, because the smithereens would probably cause more damage than the asteroid itself.'

Timing is everything. 'If you recognize that an asteroid is going to hit Earth in ten years' time, it only needs a tiny little nudge – a few centimetres a second – to steer it out of collision course. But if an asteroid is going to hit Earth a week ahead of time, you've got a big problem. I mean, an asteroid a kilometre in size – it's like a mountain. Just throwing a couple of rocks at it, or putting a rocket motor on it, is not going to move it. You've got to do something to provide a lot more momentum.'

That's when you really need to bring out the big guns. And firing a warhead from our – now hopefully redundant – nuclear arsenal towards an asteroid is a slightly more subtle affair than is painted in science fiction. 'The nuclear explosion scenarios involve exploding a nuclear charge somewhere near the asteroid and burning off some of its surface,' explains Melosh. 'The asteroid reacts to the skin burning off by moving in the opposite direction, and that way, you can give it a velocity impulse.

'The trouble is that to do that for, say, a kilometre-sized asteroid requires a nuclear charge of around ten to a hundred gigatons. Remember, the largest nuclear explosive ever detonated was fifty-eight megatons. We're talking about something tens of thousands of times bigger.'

So what's going to happen if the green lobby object to the proliferation of nuclear weapons in space, even if they might eventually save the Earth? No worries – Melosh has other plans up his sleeve. 'We were a bit worried about having these huge giga-type bombs in orbit about the Earth,' he admits. 'We felt that that kind of arrangement would probably be more dangerous to the human race than the asteroid impact itself.

'I mean, if somebody is going to make a mistake or intentionally misdirect these things onto the Earth, we feel it's important to develop alternative technologies. If we're going to deflect asteroids, we need to try to deflect them without mechanisms that potentially have a negative impact on the human race.'

Melosh has come up with a number of innovative ideas. One involves a mass driver that mines material from the asteroid and throws it off at high speed, thereby

changing its velocity and trajectory. Another scheme involves the ultimate in interplanetary billiards. 'You take a small asteroid that comes close to the big, dangerous asteroid, then deflect the small asteroid into colliding with the bigger one, knocking it off course. It's sometimes called the bankshot scenario or the pool scenario.'

What will Melosh use as the cosmic billiard cue – the force to move worlds? 'There have been people who suggest things like planting rocket motors or chemical explosives on asteroids,' he ventures, 'but the energy simply isn't enough – they're not that effective.' Instead, he proposes using solar power to the max. He intends to take a blow-torch to an asteroid.

'The scenario I've proposed with a Russian colleague, Ivan Nemchinov, is to fly a big solar collector out to the asteroid – an aluminized Mylar thing, very lightweight. Then we rendezvous with the asteroid, and focus the sunlight onto a spot on its surface. That causes the rock to vaporize – it gets really hot, and we've done experiments to verify this.

ABOVE *Earth saved! The energy of an exploding nuclear missile deflects the path of a giant asteroid hell-bent on colliding with our planet.*

'It's kind of like a rocket motor. It provides a jet, and if we keep our light focused on the asteroid's surface, we can vaporize enough of it that over a year or two we could steer it out of a collision with the Earth. We estimate that a solar collector a kilometre in diameter working five years ahead of time could deflect a one-kilometre asteroid.'

But David Hughes warns that we need more hard data about the nature of the enemy before we can draw up truly definitive plans. And as well as asteroids, he includes comets, too, in his roll-call of future cosmic threats. 'We need to know about their physics. We're not too worried about the chemistry – it's the physics that's the problem. We don't know exactly how fluffy a comet is. When it comes to an asteroid, we're not sure if we're looking at an absolutely solid body, or a collection of solid bodies just loosely held together by gravity.

'So when it comes to pushing them aside, it's important to distinguish things which are solid from those which are just piles of rubble. Otherwise, we're going to find it very difficult to deviate them. But this is for the future. We have space missions going to comets, and we have space missions going to asteroids. Only recently, we landed on the surface of an asteroid. These are questions we need answers to, and they're going to be answered in the next decade.'

Steve Ostro raises another hitch that might affect our future plans for universal domination – and eradicating the threat from space. His beloved binary asteroids, which he was so excited to discover, may present real problems for defending the Earth from impact. 'The discovery of binary asteroids has fascinating ramifications for space missions, but also for the hazard factor. Manoeuvring around one object is a hard enough job. Manoeuvring within a binary system is going to be a whole new realm of orbital dynamics that no one's really explored.

'Erik Asphaug, who's at Santa Cruz, did simulations about an impact with Castalia – a contact binary thing. He hit it with a rock travelling at fifteen kilometres a second – something like a Hiroshima bomb level of energy. And the single coherent object was completely destroyed, while the binary one wasn't.

'The impact energy just bounced off the other one, and it was left unscarred. So we discovered that the response of an object to an impact or a bomb depends on its internal configuration of cracks and density and so on. You really have to know everything about the interior to know how it's going to respond. You can't make assumptions about what it's going to be like to deflect a single body – much less a double body.

'So the impact hazard is real, and it will affect our future. We just don't know the exact timescale. But if ever there was a motive for exploring these objects, it's the possibility – which has to be taken seriously – of a one-in-a-thousand chance per century of the termination of civilization. Asteroids are both the creator and the destroyer.'

Ostro advocates learning to love the enemy. 'They're destinations for humans and for robots. They're destinations for the human spirit because humanity's long-term future, one way or another, is tied to these objects. They're also the easiest and cheapest and – in some cases – the most interesting places to go in the Solar System. The hazard they pose is a very negative kind of thing, but these objects' existence – and the fact that we discovered them when we did – is very positive. Exploration is so vital to the human spirit, and everybody is fascinated by space.'

Not just exploration. Exploitation. And Steve Ostro's radar work at Arecibo and Goldstone, in the Mojave Desert, is leading space scientists right down this route. 'One of the nice things about radar studies of asteroids,' observes Arizona University's John Lewis, 'is that we get direct information on the physical state of the surface – and sometimes, as a bonus, we get information on whether they're metallic, because the return radar signal from a metallic asteroid is enormous.'

There are several dozen metallic asteroids that are known to sweep past the Earth. One, Amun, is a solid slab of alloy two kilometres across. 'It contains more metal than all the metal that has been mined from Earth in the whole history of mankind,' notes John Lewis. 'It's easy to imagine that among the near-Earth asteroids collectively there is enough material to last the human race for millions of years.'

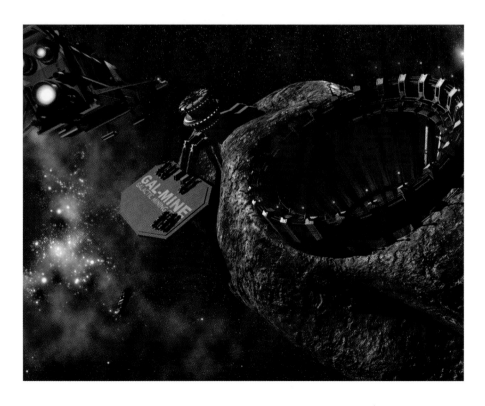

ABOVE *The future of mining? Many astronomers predict that we will extract exotic metals from asteroids over the next few centuries, and use them to build huge structures in space.*

Mining asteroids could be big business in the future. 'We wouldn't do it unless it was economically profitable,' Lewis acknowledges. 'The prospects of importing iron and nickel to Earth are, to my mind, pure fantasy. They're just not valuable enough. On the other hand, if there are any large-scale activities in space that require structural metals, it's probably a factor of a hundred to a thousand cheaper to derive those materials from asteroids than to pay the cost of lifting them from the surface of Earth.'

'Asteroids are potentially valuable in a real, dollar sense,' agrees Steve Ostro. 'They're natural resources supporting oases providing materials that will unlock the doors to our *Star Trek* fantasy – the doors that are keeping us from exploring space. Space travel is so prohibitively expensive at the moment that it costs thousands of dollars to get a pound of something into low-Earth orbit.

'We need to develop the industry – develop the capability through a combination of engineering and investment strategies that make a profitable enterprise for the private sector.'

ABOVE *One small step ... but humans may soon be making steps on asteroids. They are amongst the most accessible bodies to reach in space.*

EXTREME UNIVERSE

'I think the easiest way to make asteroid mining economically accessible is to keep humans out of it as much as possible,' adds John Lewis. 'We need a high degree of automation: the ability to operate the equipment remotely from Earth – a process called tele-operation.'

But how close is this kind of Dan Dare dream to becoming reality? Lewis believes that, with the right political will, it could all be up and running in a surprisingly short time. 'Perhaps two or three years. With that kind of notice, we could actually have such equipment available for flight. An example would be a little tank – maybe small enough to fit on the top of a desk – that could be run around on the surface of the Moon and be tele-operated from Earth.'

In an ironic twist – close to the spot where a giant asteroid blasted out a gaping hole in the Earth's surface 1.8 billion years ago – experiments in tele-operation are being conducted even now. John Lewis stands on a rubble-strewn site at Sudbury, Ontario, and gestures towards a group of pits behind him, each a couple of miles deep.

'There's a very important demonstration of tele-operation technology in the mining world here in Sudbury. The mines near which we're standing right now are being operated by equipment that is more or less conventional mining equipment, but with controls that let operators drive it remotely. There's an electronic link between the controllers and the equipment, so they can see what the equipment does, and control its every action.'

If we were to mine asteroids and exploit the cheap metals out of them, then what sort of giant engineering projects does Lewis foresee in space? What would an Isambard Kingdom Brunel of the twenty-second century regard as his or her equivalent of the SS *Great Britain* or the Great Western Railway? 'If there were a low-cost source of metals in space, all kinds of plans that have been on the back-burner for decades would move to the fore,' maintains Lewis. 'The first example would be solar-power satellites – large arrays of solar cells in orbit around the Earth that would capture sunlight, convert it into radio-frequency energy, and beam it down to receiving antennas on the Earth. Then you'd convert back that microwave power into electricity, and put it into the national grid.

'This scheme has been well studied, but the principal difficulty is the economic cost of launching these satellites from the surface of Earth. On the other hand, the overwhelming majority of the mass of these satellites is in very simple low-tech stuff – it's beams and girders, it's nuts and bolts, it's wires. Things of that sort could be fabricated by an automated plant in space, out of materials that are native to space, saving the enormous launch cost from Earth.'

Materials mined from asteroids may even help the human race itself to move out into space – perversely, to escape the asteroid threat. 'We're going to have to become space travellers to escape the hazard,' warns Steve Ostro. 'And I think we'll need technology that's beyond what anybody has thought of.'

But the first glimmers of that technology are beginning to shine through, as Ostro relates. 'In 1998 an object was discovered just a week or two before it made a very close approach to the Earth – just twice as far as the Moon. And 1998 KY26 happened to be a very small object, only thirty metres across. That's the size of an American baseball diamond, or a typical small building, and it rotates once every eleven minutes or so. And it's the easiest place to go in the Solar System, and the cheapest place to go right now.

'It's made of a very primitive kind of material called carbonaceous chondrite that contains maybe ten or twenty per cent water. We could go there and extract that water for life support. Imagine I was holding up a chunk of the stuff, a little larger than a soccer ball – I could heat it up and get enough water to live off for a day. 1998 KY26 contains the equivalent of a million of those balls, so that object represents survival rations for a million people for one day, or one person for a million days.

'Everything we need in space, all the resources, are in the carbonaceous chondrite asteroids. This is the first one we've discovered that's very accessible, and it's inevitable we're going to go to this object. Within the next five years, maybe there'll be a fly-by and after the rendezvous mission, a lander mission.

'And at that point we'll have enough information where we can start to do experiments on actually extracting the water. Maybe within twenty years we could have a human mission to this object. Then we'll know enough where we can really start to have fun with it.'

Ostro proposes populating the asteroid with an army of robots that could extract the life-support materials – hydrogen and oxygen, in addition to water itself – so that it becomes a veritable oasis in space. 'Then a team of astronauts goes there and spends five years living off the asteroid. After that, you cover it in robots and hollow it out.'

Living inside an asteroid wouldn't be such an uncomfortable experience, Ostro believes. 'It would be much roomier and more massive than the International Space Station. Plus, the shielding provided by the shell of the object from the lethal effect of cosmic rays in space is much thicker. And we would have everything we need to live. The organics on KY26 exist in a concentration of probably several per cent by weight. People have done experiments where they've grown bacteria in carbonaceous chondritic soil to demonstrate the nutrient value and potential for farming, using this material as potting soil.

'Then one day, we'll get to the point – perhaps in this century – when there are people conceived on the object. They'll be the first real extraterrestrials – and we'll have completed the process of turning an asteroid into a womb.'

Our future relationship with asteroids could go still further than this. Even more than a womb, asteroids could be ultimately used as the celestial equivalent of Noah's Ark: giant colonies where people are born, live and die, where the human race could be protected from threat for ever. But unlike their terrestrial predecessor, these space-arks

OPPOSITE *A vast space ark – a hollowed-out asteroid, populated with generations of spacefarers living on board – continues its tour of inspection of our Galaxy.*

would be constantly on the move – taking years, millennia even, to explore the hidden bays and headlands of our Galaxy. And the key to making it all happen lies within the asteroids themselves.

'You get the water, you get the hydrogen, you get the oxygen, and you can use it to make fuel,' explains Ostro. Not only do asteroids come fully formed with their complement of potting compost and water, they also come kitted out with their own onboard supply of almost limitless rocket propellant.

This underpins John Lewis' dream that asteroids – far from being the vermin of space and the bringer of mass extinctions – might be the saviour of life on this planet. 'H. G. Wells thought about this many years ago, and he said the choice is the Universe or nothing. We can elect to have a future where we limit ourselves to being trapped on the surface of a planet in a gravity well, in an environment with limited resources and beset by natural calamities. I don't think I want that kind of future for my descendants.

'There are quite a few people out there who feel the colonization of space is inevitable. As a student of human affairs I think nothing is inevitable. But I think it will happen if it's economically profitable – if there are niches out there that human beings can occupy and prosper in. The resources in the near-Earth asteroids are comparable to what it takes to run a civilization of ten billion people.'

Steve Ostro believes that space colonization, and even travelling the Galaxy in asteroids, *is* inevitable. And asteroids are a vital part of that process. Ultimately he sees

ABOVE *As well as extracting the metals from asteroids, we'll be able to use the water from comets for making the constituents of rocket fuel. Saturn's icy rings will be an excellent source, too.*

them not as our enemy, but as part of our future. 'These objects are small worlds, yes they are,' he says, firmly. 'And some day, people are going to live on them – even in them.

'For those people, these objects are going to be their world. And they're going to talk to humans who stay on Earth, and they're going to ask them, "You're a big huge planet – do you think of yourself as a world?" '

ABOVE *Site of starbirth: the hauntingly-beautiful columns of the Eagle Nebula. Asteroids carrying humans may venture here one day – and we will have made our peace with the enemy.*

Violent

chapter seven

Universe

It's late June 2001, and **American scientist David Spergel** is an excited man. Within months of the birth of his third child, he's about to take the ultimate baby pic: a snapshot of the Universe just after its birth.

Instead of the local maternity ward, Spergel is off to Cape Canaveral in Florida – hallowed ground for space explorers. Next to the giant launch-pads of the Kennedy Space Center, where astronauts blasted off to the Moon and the space shuttles now fly to the International Space Station, the Cape Canaveral complex bristles with smaller gantries. Nestled into one of them is an unmanned Delta rocket; and, in a few days, David Spergel's dreams will ride its fiery ascent from planet Earth.

'The only experience I've had close to this is the birth of a child,' he enthuses, 'in that you're there at the conception, excited about the result, and now I'm waiting for the experts over here to perform the operation, to see what comes out.' What comes out for Spergel may answer the basic questions about the Universe itself.

'When I get down to the Cape,' he continues, 'I think the launch itself is going to be very exciting – but it's also going to be a little scary. I mean we've invested many years in making this work. The folks at NASA have done everything one could to assure a successful mission. But it's inherently a risky thing, to put a sensitive satellite on a big piece of explosive.'

On the day, Saturday 30 June, the explosive power of the Delta is successfully controlled. Lifting slowly at first, then accelerating into the cloudless Florida skies, the rocket lofts Spergel's satellite – safely – into orbit.

ABOVE *Keeping station far out beyond the orbit of Earth and Moon, the MAP spacecraft is set to deliver pictures of the infant Universe soon after the Big Bang.*

OPPOSITE *Giant magnetic loops on the Sun can explode as life-threatening solar flares; but they are puny on the overall scale of cosmic violence.*

Watching with baby Joshua strapped to his chest, Spergel is close to tears. 'That felt great; I had this tremendous sense of relief. I mean it was really exciting to see it come off – it was just a beautiful thing to see.'

Spergel's eye on the baby Universe is called *MAP* – the *Microwave Anisotropy Probe*. With scientists' fondness for punning, it's no coincidence that *MAP*'s task is to map the early Universe. It will spy out radiation that's been travelling across space from the very earliest days of the Cosmos. The result will be a detailed picture of the hot gases that erupted from the Big Bang itself.

MAP has a long way to go before it gets down to business. It needs a quiet environment, well away from the pollution of stray radio waves from transmitters on Earth. After swinging around the Moon, *MAP*'s destination – three months later – is a point a million miles from our planet, on the side away from the Sun.

MAP is station-keeping far further out into space than humans have travelled – four times further off than the Moon, where astronauts walked in the years 1969 to 1972. But even this is a trifling 'small step' on the scale of the entire Universe: the radiation that *MAP*'s observing has come from regions a billion billion times more distant still.

Compared to our home on Earth, the Moon was a pretty inhospitable place to visit – 'magnificent desolation', in the words of Buzz Aldrin, the second man on the Moon. But as we travel further and further out into space – on the trail of the Big Bang – we find a Universe that's more and more violent.

Our biggest planetary neighbour, Jupiter, has an uncomfortable habit of swinging asteroids our way, to impact our planet. Comets from the outer regions of the Solar System can likewise hurtle our way, to the detriment of life on Earth.

Even the Sun, which provides us with life-sustaining heat and light, is a fickle protector. Giant explosions on our local star, and violent storms in its atmosphere, both send out deadly bursts of radiation. Fortunately for life on Earth, our planet's magnetic field provides an invisible protective veil. Our only hint of the Sun's ferocity is the glowing coloured curtains of the aurorae – the Northern and Southern Lights – where the air above our heads mops up the final vestiges of a solar attack.

The Sun – as far as we know – is a run-of-the-mill star. Some of its brethren pose a far more dangerous threat to any planets they may possess. These stars are prone to superflares: giant eruptions lasting anything from an hour to a whole week.

'We found that superflares do occur on disturbingly normal solar-type stars,' says American star expert Brad Schaefer, who reported his discovery of superflares on nine Sun-like stars in 1999. 'Superflares are a hundred to ten million times more energetic than the brightest solar flares.'

If the Sun emitted a superflare, it would parch the Earth for months. More dangerously, its radiation would destroy our ozone layer, letting the Sun's lethal ultraviolet penetrate down to sea-level. 'The loss of ozone could kill the food chain and probably cause mass extinctions,' Schaefer calculates.

Fortunately for us, there aren't any signs that the Sun has attacked us in this way in the past – perhaps a good omen for the future.

Beyond the Sun's immediate family, the vast reaches of space stretch out, beyond anything that the mind can understand. To comprehend what lies beyond, even astronomers sometimes take refuge in analogies, and resort to mental models that scale the Cosmos down to a manageable size.

'My favourite analogy,' says astronomer Brian Boyle, 'is to say that if the Sun were the size of a six-inch balloon, the nearest star would be four thousand miles away. That gives you an idea of the emptiness of space. And this star – Proxima Centauri – is right next door in cosmic terms.'

Scottish by background, Boyle now observes the Universe from the opposite side of the world. Instead of rounded Highland mountains, he works in a wonderland of sheer volcanic stumps and jagged walls of solidified lava – one aptly named the Breadknife – that makes up the Warrumbungle region of New South Wales.

Perched on one of these mountains is the Anglo-Australian Telescope, a colossus of steel and glass that reflects the skies with a precision mirror almost thirteen feet across. 'This telescope was built over twenty-five years ago,' Boyle explains, 'and it's still one of Britain's premier astronomical instruments. It's done so by always having the latest in technology, the latest in instruments and devices that can capture the light from distant stars and galaxies.'

For the Anglo-Australian Telescope, our cosmic neighbour Proxima Centauri is one of the less interesting kids on our own cosmic block. This star, along with the Sun and most of the stars you can see for yourself in the night sky, live in a quiet region of our own 'star-city', the Milky Way Galaxy.

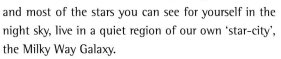

It holds far raunchier places than our local backwater. Among the hundreds of billions of stars making up the Milky Way, we encounter the traumas of star-birth, deadly encounters between stars – and the spectacular and violent suicide of stars at the end of their long lives.

To navigate the expansive highways and byways of the Milky Way, astronomers have to abandon the puny measurements of miles or kilometres. Instead, they turn to the shining path trodden by rays of light. A beam of light moves at a phenomenal speed: fast enough to girdle the Earth seven times in a single second. Interstellar space challenges even a speeding light-ray. The light from the star Proxima Centauri has a four-year journey to undertake before it reaches us.

'So Proxima Centauri is four light years away,' continues Boyle, 'but that's really right next door in cosmic terms. The centre of the Milky Way is twenty-four thousand light years away.'

The Anglo-Australian Telescope has opened the eyes of the human race to much of the Milky Way's magnificent scenes of creation and destruction. At a time when most astronomers were content to take black-and-white pictures, an English photographic chemist, David Malin, began to use the telescope in the Warrumbungles to methodically reveal the wonders of deep space in glowing colour. It was as revolutionary as the replacement of black-and-white television by colour sets.

'We can see the whole life-cycle of stars,' says Brian Boyle. 'We see clouds of gas and dust in which new stars will be born; we see young stars being born; we see old stars die; we see old gaseous envelopes from stars that have just died.'

ABOVE *The Anglo-Australian Telescope investigates the depths of space through the slit atop this huge silo-shaped building, towering over the eucalyptus trees.*

The glowing nurseries of baby stars have acquired fanciful names: the Eagle Nebula, the Lagoon, the Trifid. In their depths, young stars suffer severe birth-pangs. Newly born stars flex their muscles as they coalesce from clouds of gas. The biggest baby stars are cosmic cuckoos. They emit energetic radiation and shoot out piercing jets of gas that damage – and even destroy – the other fledglings in the interstellar nest.

At the other end of the cosmic lifespan, the Anglo-Australian Telescope has revealed the colourful wraiths of gas from stars making their exit from the celestial stage. The Helix Nebula, the Bowtie, the Owl, even the Ghost of Saturn – names that mirror the intricate shapes woven by dying stars. One day, our Sun will end in this way: and what fanciful name will its dying gases attract?

ABOVE *Under clear dark skies, the Milky Way is a staggering sight. This shining band is composed of hundreds of billions of stars.*

OPPOSITE *The Rosette Nebula is a nursery of hot young stars, surrounded by tatters of glowing gas, in this image from pioneering astrophotographer David Malin.*

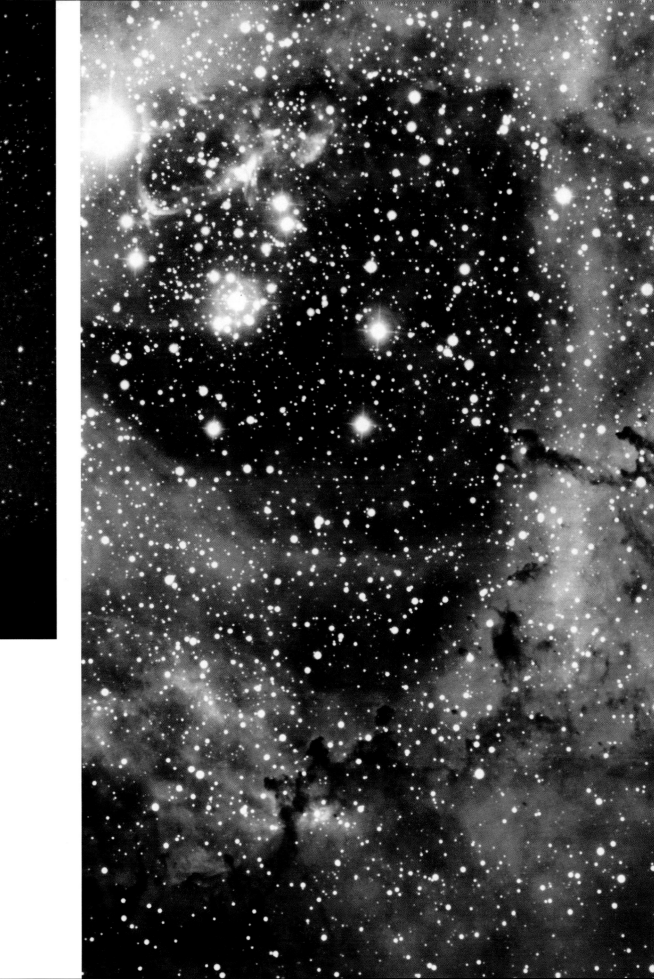

Fortunately for us, space is big enough for us to spectate on these scenes of cosmic violence without becoming involved. When it comes to the ultimate in stellar suicides, though, nowhere in the Galaxy is safe.

The alarm bells rang, quite literally, on 27 August 1998. A powerful burst of radiation from the depths of the Milky Way sped past the Earth, triggering an alert on dozens of spacecraft all over the Solar System – from Earth orbit out to a probe near Jupiter. One even shut itself down because of the overload. As the radiation reached the night-side of Earth, it lit up the upper levels of our atmosphere to their daytime brightness.

Astronomers quickly homed in on the culprit. It was the corpse of a dead star, invisible to ordinary telescopes, but still powerful beyond the grave. In its death throes, the star had blown away its outer layers. The thick core at its heart had shrunk down to a dense ball, seething with magnetic power.

A star corpse like this is a cosmic zombie, wielding powers that the star never possessed during its lifetime. Star corpses go under various exotic names – white dwarfs, neutron stars, pulsars – but the variety that flexed its muscles on that occasion was a magnetar. Immense magnetic power spills out of its tiny globe, a million billion times stronger than the Earth's magnetic field. If you were rash enough to land on a magnetar, its magnetism would kill you – by rearranging the atoms in your body.

ABOVE *When the Sun reaches the end of its life, its gases will billow out into a colourful nebula – like the Egg Nebula seen here – and, in the process, destroy the Earth.*

Suppose a magnetar were to pass the Earth at the Moon's distance. Its magnetic force would erase the information on your credit card, and suck metal pens out of your pocket and lift them into space. As it passed, your credit cards or pens would be the last things on your mind: even at the distance of the Moon, the magnetar's radiation would kill you in seconds.

And even if it's sited in the further reaches of the Milky Way, a magnetar can extend its malign influence as far as our planet. Superflares on an ordinary star may be worrying some astronomers, but they pale into insignificance compared to a magnetar's flare; in comparison, it's like the igniter spark of a kitchen stove versus a lightning strike. When a magnetar unleashes the pent-up energy of its magnetic field, it's an explosion billions of times more powerful than the worst superflare ever seen. That's the cosmic fury that the Earth experienced in August 1998.

A magnetar is only a few miles across – and yet it contains more matter than the Sun. It's so dense that a mere pinhead of its material would weigh more than a fully-laden oil supertanker. How does an ordinary star end up as a compressed ball of malevolent energy?

The answer involves an explosion of unparalleled violence: a supernova. It may last only a short time – compared to the lingering malice of the resulting magnetar – but during this immense eruption the dying star can outshine the entire galaxy of stars where it lives. Here, we are ramping up the scale of extraterrestrial violence by another notch . . .

Such star explosions are rare, and it's been centuries since anyone has seen a supernova explode in the Milky Way. But supernovae are so brilliant that we can see them across vast reaches of space, in other galaxies way off beyond the Milky Way – as three very different astronomers were to discover one fateful night in February 1987.

'I was taking this photographic plate out of the fixer,' recalls Canadian astronomer Ian Shelton, 'to take a quick look. I usually do this before I finish processing, to make sure the brightest stars appear as nice round star images – that makes me feel good. The images were excellent, beautiful. I took a few moments to marvel at the galaxy's bar; and then gravitated over to the gaseous tentacles of this giant nebula in the galaxy.'

Shelton's photographic subject that night was the Large Magellanic Cloud, a patch of glowing light in the southern skies. Along with a smaller luminous 'cloud', it had piqued the curiosity of Ferdinand Magellan on his great circumnavigation of the globe in 1521. After Magellan died en route, the expedition's official recorder suggested the two strange celestial sights should be named in his honour – the Large and Small Magellanic Clouds.

It took several more centuries before astronomers established that these 'clouds' were galaxies of billions of stars, orbiting around our Milky Way just as the Moon

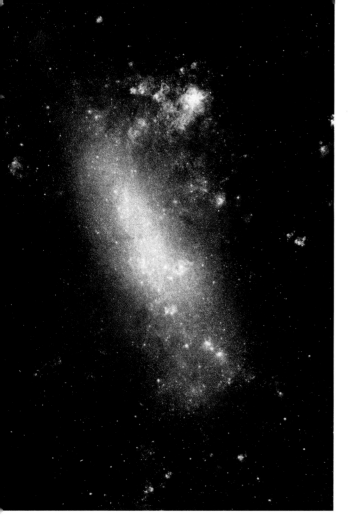

circles the Earth. Through a telescope, the Large Magellanic Cloud is a treasure trove of coloured stars, clusters of stars swarming together, and filamentary nebulae – with much of the action concentrated in a central straight 'bar' of stars.

Sadly for northern-hemisphere astronomers, this cosmic cornucopia is forever hidden below the horizon. Ian Shelton had travelled from Canada to an observatory in Chile to investigate its stars. The whole galaxy was as familiar as it was fascinating to him. And as Shelton's gaze moved away from the galaxy's central bar to the adjacent nebula, on that fateful night in February 1987, he realized something was awfully odd.

'There was a star right beside the nebula,' he says, 'and it's a bright star. Then I caught on: there isn't a star that bright in the Large Magellanic Cloud! So at this point I was starting to get either excited or worried: is it a problem with the plate, or is it real?'

Shelton left his telescope dome and went outside. 'I walked up the path a bit, waited for my eyes to dark-adapt – and then looked up. And sure enough there was this star just sitting in the Large Magellanic Cloud, plain as day – or as plain as night!'

A few hours earlier, another pair of human eyeballs had seen the new star. Oscar Duhalde was on duty at the same observatory that night. Though he worked as a technician, not an astronomer, his lifetime under the clear Chilean skies had left a deep impression. When Shelton mentioned the new star, Duhalde reported that he'd seen it already – but hadn't realized it might be important for the astronomers.

Meanwhile, far from the international professional observatory sited expensively in the Chilean mountains, an amateur astronomer was setting up his backyard telescope in New Zealand. Albert Jones was a retired miller, now devoting his nights to checking out stars which are inconstant in their brightness.

'I turned my telescope towards the Large Magellanic Cloud,' he remembers, 'and there was something in the field of view that wasn't there the night before. Goodness gracious me, I thought, "This is something unusual – a bright blue star."'

ABOVE *The Large Magellanic Cloud is a smaller sibling to our Milky Way Galaxy, full of young stars and bright glowing gas clouds.*

OPPOSITE *Supernova 1987A, before and after! The arrow (left) points to the doomed star in an old photograph; in February 1987 it exploded in a blaze of glory (right).*

'I quickly marked it on a star-chart,' Jones continues. 'Then I was going to make an estimate of its brightness, but the clouds beat me to it.' After a frustrating half-hour, the clouds rolled away again. Jones checked the star's brightness, and put a phone call through to a colleague, who passed the news on to the nearest world-class observatory – the Anglo-Australian Observatory in New South Wales.

'One of the Anglo-Australian Telescope's great achievements,' says Brian Boyle, its present director, 'was that it was the first major telescope to follow up in detail Supernova 1987A.'

From the catwalk that girdles the telescope dome, the new star was clearly visible low down in the south. But – frustratingly for the astronomers there – Supernova 1987A was too far down. The telescope had a sophisticated computer-controlled drive that prevented it from tilting too near the horizontal, in case it accidentally hit something. Determined not to miss the scientific opportunity of a lifetime, the then-director installed volunteers to watch the telescope's every move, as he overrode the computer safeguards and swung the great eye-on-the-sky towards the exploding star.

Over the following weeks, the exploding star in our neighbouring galaxy grew steadily brighter. At its peak, the supernova was shining 250 million times more brilliantly than our Sun. Its light was literally blinding for the giant telescope and its electronic light detectors, more used to straining at the faintest and most distant objects in the Universe.

But it was also a unique chance to spread out a supernova's light and examine the spectrum of light in unprecedented detail. Following the observatory's tradition of

innovative instruments, optical expert Peter Gillingham hastily assembled the world's most powerful spectrograph: there was no time for carefully machining the instrument from metal, so he crafted this unique scientific instrument from wood!

Supernova 1987A taught astronomers why stars explode. In its final moments, the core of a giant star collapses on itself. In this inferno, the temperature soars to an unimaginable 50 billion degrees. It releases a flood of tiny particles called neutrinos, which blow the rest of the star apart.

This explosion seethes with nuclear reactions, transmuting the exploding gas. The radioactive atoms light up the expanding debris and keep it shining brilliantly for months on end.

Supernova 1987A was that lucky chance: a rare event that happens to occur right on our cosmic doorstep. But the huge eye of Anglo-Australian Telescope allows it to cast its net far wider. Gazing out over billions of light years, it has investigated hundreds of supernovae in much more distant galaxies.

And out here – the broadest cosmic stage – astronomers have discovered a still rarer kind of star blast. Once again, we are racking up the level of cosmic violence: a hypernova is the biggest explosion since the Big Bang itself.

Outbursts from hypernovae were first detected back in the 1960s, though no one knew what they were. Indeed, they were not even discovered by astronomers. US Department of Defense satellites were checking for nuclear explosions that violated the Test Ban Treaty when they found bursts of radiation coming not from clandestine tests on Earth, but from far off in space.

For decades, these violent outbursts ranked among the biggest puzzles in astronomy. Some astronomers thought they might be a merely local disturbance, a new kind of outburst just on the edge of the Solar System. Or were they out in the Milky Way? Perhaps, again, they came from the very depths of space.

The puzzle had no immediate solution, because astronomers were always being caught on the hop. By the time they'd got the data from satellites monitoring gamma rays, and worked out where in the sky they should look, the outburst had faded away. It was time to move beyond the human astronomer.

In January 1999, a celestial burst triggered alarms on NASA's giant Compton Gamma Ray Observatory, in orbit around the Earth. Without human intervention, the message was beamed to New Mexico, where a robot telescope swung round to that region of the sky. Within a minute of the burst hitting the satellite, the automatic telescope had picked out a corresponding flash of light. If anyone had been looking, they would have seen a temporary star bright enough to be seen with binoculars.

The New Mexico telescope provided astronomers with a precise position to look. When the powerful Hubble Space Telescope swung round to this spot in the sky, it found the final faint afterglow of the explosion. The detonation had occurred in a galaxy billions of light years away from the Earth.

It was the long-sought proof. These dramatic outbursts are star explosions even more extreme than a supernova – hence the new name hypernova. To appear so bright in our skies despite its immense distance, a hypernova must shine – for a brief period – more brilliantly than everything else in the Universe put together.

A hypernova exploding in the Milky Way would be lethal for life on Earth. Its radiation would kill anyone who didn't take shelter deep underground. And even that wouldn't provide long-term security, for the radiation would also destroy Earth's atmosphere . . .

For all its ferocity, a hypernova is a short-lived catastrophe. Out here in the realm of the distant galaxies, billions of light years from home, astronomers have found an even more energetic beast in the cosmic jungle. Though it never peaks at the supreme brilliance of a hypernova, a quasar is an even more fearful cosmic predator, with its radiation constantly at danger level.

It was in the early 1960s when astronomers first came face to face with the quasars. Until then, they had thought of the Universe as a fairly tame place. Placid stars whirled languidly around in picturesque galaxies, out as far as telescopes could see. At this time, the Big Bang was only a theory; other manifestations of the violent Universe – magnetars, pulsars, neutron stars, black holes – were only ideas in the minds of a few other-worldly theorists.

And quasars didn't immediately rock the boat. They seemed to be stars that broadcast radio waves, bringing them to the attention of the newly invented radio telescopes, including the giant dish at Jodrell Bank. At first they were called 'quasi-stellar radio sources' – later mercifully abbreviated to quasars!

In 1963, Dutch-American astronomer Maarten Schmidt analysed the light from one quasar in detail, and realized it wasn't a star in the Milky Way but something lying billions of light years from us. It was a staggering discovery. This quasar must be shining with the brilliance of hundreds of galaxies like the Milky Way, yet it was not much bigger than the orbits of the planets around the Sun. How could so much energy be so closely packed?

'What we think now,' says Brian Boyle from the Anglo-Australian Observatory, 'is that there's a massive black hole deep in the heart of some galaxies. It's a black hole that probably weighs about a hundred million times the mass of the Sun. And it's eating the centre of the galaxy out.'

A black hole is Nature's ultimate abyss. It's a region where gravity reigns supreme. From the outside, a black hole looks like a dark silhouette against the distant stars. But it's not made of anything solid – it is literally a hole in space. Once you fall into a black hole you can never get out again. It's black because even light cannot escape from this bottomless pit.

'What this black hole is doing,' Boyle continues, 'is converting the mass of the stuff that it's eating into energy. Now most of this disappears down the throat of the black hole, but a little bit of this energy manages to escape just before it disappears into the black hole. It's the remnants of this wonderful meal that the quasar is enjoying – a gigantic cosmic burp.'

These burps are dangerous. The quasar belches out radiation of all kinds: X-rays, radio waves, gamma rays and intensely brilliant light. 'The light from this tremendous cosmic feast can outshine all the stars in the host galaxy by a hundred or even a thousand times,' says Boyle. 'So these are massive inter-galactic beacons that outshine ordinary galaxies, and we can see them to much further distances – maybe ten thousand million light years.'

Boyle's enthusiasm for quasars reflects the latest 'first' for the Anglo-Australian Telescope: the biggest ever census of quasars. 'Previous surveys of quasars have been

ABOVE *The awesome power of a quasar is revealed in this artist's impression: residing in the core of a distant galaxy, the quasar spits out gas, light and dangerous radiation.*

OPPOSITE *At the focus of the giant Anglo-Australian Telescope, these tiny glass fibres are arranged so that each picks off the faint light from a different galaxy or quasar.*

quite small,' he explains, 'typically comprising only a few hundred or a thousand objects. That simply wasn't good enough to give us a good census of the Universe. It's a bit like conducting a census of the British population, for an electoral poll. You wouldn't just survey a few hundred people in a particular constituency, say Surbiton, to get a valid picture of what's going on. You need the big picture – and that's why we embarked on this large-scale survey of the Universe.'

And it's given a new lease of life to the venerable telescope. Once again, its original mirror is focusing light on an innovative piece of equipment. While most big telescopes now are tunnel-visioned giants, the Anglo-Australian Telescope has broadened its view so it can encompass a region of sky two degrees across – four times the diameter of the Full Moon.

'The two-degree field, or 2dF,' enthuses Boyle, 'is a clever piece of equipment that allows you to study in detail the light from not just one galaxy or quasar at a time, but up to four hundred at once. So, in an era when the largest telescopes on our planet are four to six times as large as the Anglo-Australian Telescope, by observing many objects simultaneously we can still be competing effectively with these other behemoths.'

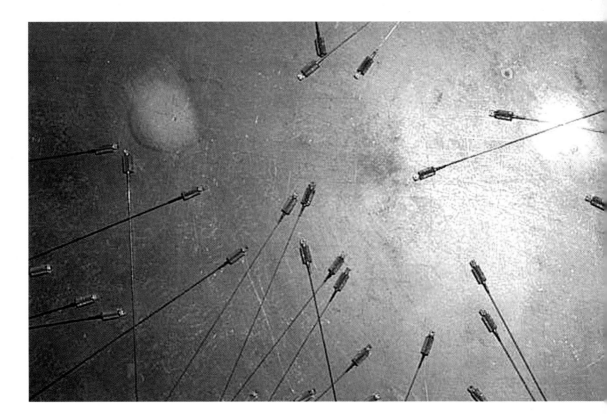

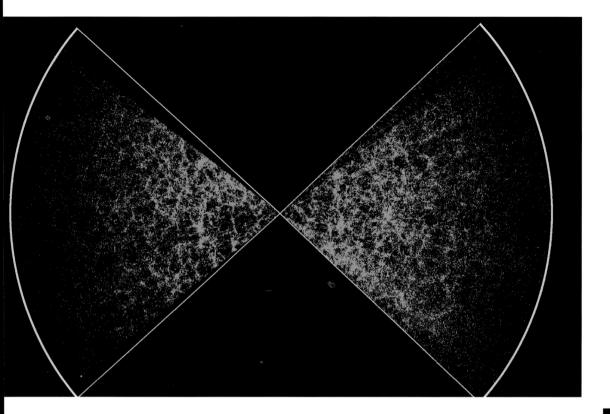

The cosmic census has already passed any previous survey of the depths of the heavens. By the middle of 2001, the 2dF survey had netted a staggering 175,000 galaxies and twenty thousand quasars – and the number is set to grow. But for astronomers, this is not just stamp-collecting writ large. They are looking forward to the science to be extracted from the raw statistics.

'We're looking well beyond our cosmic backyard,' says Boyle. 'For example, the nearest cluster of galaxies to the Milky Way is the Virgo Cluster – and that's sixty million light years away. But even this is just a speck in terms of cosmic scales. I sometimes liken the Universe to a room thirteen metres across, and on that scale the Virgo Cluster is only a centimetre away from us: the rest of the room is *terra incognita* – the realm of the 2dF.'

Size is not all that's important: the survey has – quite literally – an extra dimension, the dimension of time. Instead of an instant snapshot, like an election census of Britain, 2dF is providing an in-depth view of the heavens in time. We are seeing the population of the Universe not just as it is now, but as it was long ago.

'You have to think of astronomy as a kind of cosmic palaeontology,' says Boyle. 'The same way as a palaeontologist digs down through the fossil records of Earth to

ABOVE *A grand plan of the Universe appears in regions covered by the 2dF survey, five billion light years across with the Milky Way at the centre. Each dot is a galaxy.*

OPPOSITE *Most of the galaxies in Stephan's quintet lie 350 million light years away: we see them as they were long ago, when life on Earth first colonized dry land.*

piece together the evolution of life on our planet, so when astronomers use telescopes to look out into the Universe they're looking back in time – because of the amount of time that light takes to travel to Earth from these distant objects.'

Gaze up at the sky, and you're looking directly at history. The light that's entering your eyeball now has been travelling through space for some time before reaching you. Even light from the Sun takes eight minutes to reach us. If some evil alien were to switch off the Sun at this instant, you'd be able to finish your cup of coffee before you'd even know it's gone.

We see the nearest star, Proxima Centauri not as it is now, but as it was four years ago. And when astonished astronomers first sighted Supernova 1987A, in the Large Magellanic Cloud outside our Milky Way, they were seeing an explosion that had actually happened 170,000 years ago. Since the time of the Old Stone Age, the light from this celestial funeral pyre had been winging its way towards us.

In the intervening years, many other stars must have died in the Large Magellanic Cloud. But we still see them as they were in the prime of life: the bad tidings have yet to reach the Earth.

And the larger the telescope at your disposal, the more powerful a time machine you control. 'With a big optical telescope like the Anglo-Australian,' Boyle explains, 'we look back about thirteen thousand million years into the past. That's about ninety per cent of the way back in time to the beginning of the Universe.'

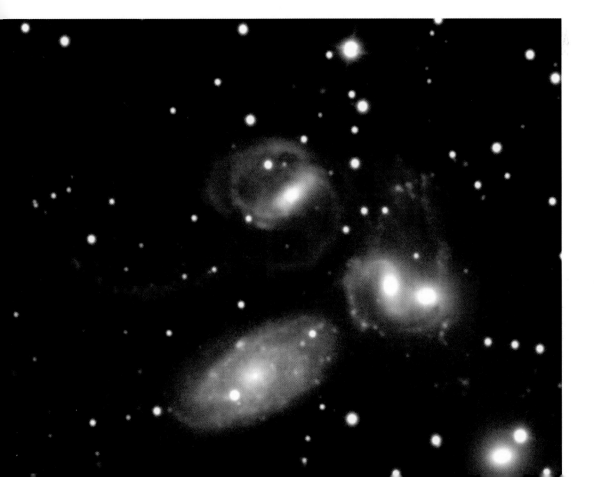

Dig down through the Earth's rocks nine-tenths of the way back to the birth of our planet, and you'd expect to find rather different fossils from the creatures on Earth today. And so it is with the Universe.

'In our "cosmic palaeontology",' says Boyle, 'we're digging out strata not in terms of layers of rock, but time strata in distances out to these remote objects.' And so the 2dF is an ambitious quest to chart both the geography and the history of the Cosmos.

'We can only see galaxies in large numbers out to a distance of three billion light years from Earth,' he continues. 'Now that was a long time after the Big Bang, so there's a large gap there – a big bit missing from the fossil record, if you like. It's like a palaeontologist only being able to dig down as far the Cretaceous–Tertiary boundary: you wouldn't know anything about dinosaurs or anything about the early evolution of life on Earth.

'Similarly, we don't really know much about the early evolution of the Universe, and in particular the large-scale structure of the Universe, the fabric of the Universe, if you like. So our goal was to use these quasars – these bright objects that can be seen to great distances – to fill in the missing link in the evolutionary history of the Universe.'

Boyle's survey isn't the first to penetrate this far into the Universe. In December 1995, the Hubble Space Telescope scrutinized a small patch of sky in the region of the

ABOVE LEFT *From its lofty perch above Earth's distorting atmosphere, the Hubble Space Telescope has the sharpest views of the distant Universe.*

ABOVE RIGHT *The twin Keck Telescopes on Hawaii complement the space telescope: while their view is more blurred, these giant mirrors can see galaxies thirty times fainter than the much-smaller Hubble.*

Plough, staring at the same spot for ten days to record the faintest and most distant galaxies and quasars.

'People do say, "Why don't you use the distant galaxies you see in the Hubble Deep Field?" ' explains Boyle, 'but it's terribly limited in view, and not sampling a fair volume of the Universe. Here's an analogy. It's like looking through a keyhole into a room and trying to work out the pattern on the wallpaper on the wall beyond. You don't know if the narrow limited view you've got is allowing you to see a typical pattern of the wallpaper – especially if the pattern is larger than the area you can see. And that's essentially what the problem was before: the pattern of the Universe was larger than the area we were looking at.'

The broad view of the 2dF survey has now started to pick out the pattern in the cosmic wallpaper, the scattering of brilliant erupting quasars that exist far out in space and far back in time. And they are not sprinkled at random. Instead, there are giant regions where Boyle finds more quasars, and regions where there's a dearth.

'We are detecting structures that are several hundred million light years across,' he reveals. 'What it means is that very early on, there had to be significant concentrations of matter in order for quasars to form.'

But how exactly were these brilliant beacons ignited? And what act of cosmic genesis led to the birth of the galaxies around them?

To answer these questions, we must turn from the pattern of the cosmic wallpaper to its individual spots. And here the keyhole approach of the Hubble Space Telescope comes into its own. From its vantage point above Earth's turbulent atmosphere, Hubble has a singularly sharp view of the distant Universe and its infant galaxies.

But Hubble is tiny compared to the behemoths that have now been built on Earth's surface. While none can match its views for sharpness, the giant ground-based telescopes can catch the light from fainter galaxies, and analyse it in finer detail. The two heavyweight champions in the astronomical league are based on high extinct volcanoes, one high above the beaches of Hawaii and the other in the Chilean desert.

Like a pair of disembodied eyeballs, the twin domes of the Keck Observatory house the biggest single telescopes in the world. Each of the two Keck Telescopes has a giant mirror over thirty feet across. That's too big for even the most ambitious astronomer to make and then transport up a mountain 14,000 feet high. So the telescope's mirrors are made of dozens of panels, fitted together so precisely you can't see the join.

Over in Chile, European astronomers have pushed technology to the limit, with a quartet of telescopes each boasting a mirror twenty-five feet in diameter. Though each telescope is smaller than either of the Kecks, they can be ganged together to create the world's most powerful astronomical instrument – appropriately called the Very Large

Telescope. If that sounds too prosaic, the four individual telescopes have their own names, taken from the local Mapuche language. At the suggestion of a Chilean schoolgirl, these eyes on the sky are Antu (the Sun), Kueyen (the Moon), Melipal (the Southern Cross) and Yepun (Venus).

Hubble, Keck and the Chilean quartet have probed individual galaxies in their youth – some with quasars erupting at their cores, some living more quiescent lives. The great time machines have shown that young galaxies are smaller and dimmer than the Milky Way where we live. Brian Boyle suggests: 'Perhaps the big bright galaxies we see at the present day have been assembled from these little bits of galaxies.'

But the quasars within these scrappy little galaxies were blazing more fiercely than anything we have around us today. 'What may be happening,' says Boyle, returning to his cosmic feast analogy, 'is that the feeding efficiency for quasars in the early Universe was much more effective. The black holes have stayed in the galaxies; it's just that the quasars are being starved as they come to the present day.'

ABOVE *The aptly named Very Large Telescope in Chile combines the light from four different telescopes to create the world's most powerful eye on the Universe.*

The Age of the Quasars would have been an exciting – and dangerous – time to live. The Universe itself was only a fraction of its present size, with galaxies crowded in on each other. Beyond the bright young stars in our night sky, we'd be treated to the sight of dozens of luminous clouds, small galaxies like the Magellanic Clouds today. Within these galaxies – and beyond, in galaxies too faint to be seen with the naked eye – are brilliant blue points of light. Many of these quasars are squirting out luminous jets, which stream millions of light years into space. If such a beam of pure energy happens to be heading our way, we won't be around for long . . .

But even this violent epoch pales into insignificance when compared to the Universe a billion years earlier. Then, the Universe itself was born in the ultimate act of violence – the Big Bang. Astronomers hit a snag, though, when they try to peer back beyond the Age of the Quasars. Through even the world's biggest telescopes, there's just nothing to be seen.

'The period that existed further back than you can see with an optical telescope is called the Dark Ages,' says British cosmologist Richard Saunders, 'and we don't know any of the details of what went on then.' But that period of ignorance doesn't worry Saunders too much. He's determined to leapfrog the Dark Ages, and look for answers even earlier in time.

This quest has brought him to the Canary Islands: while most people go there to feel the warmth of the Sun, Saunders commutes regularly from Cambridge to Tenerife to catch the heat of the Big Bang. And he's surveying the uttermost depths of the Universe with an instrument that bears no family resemblance to the giant optical telescopes in Australia, Hawaii and Chile. 'Our instrument can look back further than any of the "ordinary" telescopes – including the Hubble,' he declares proudly, 'and that feels pretty interesting.'

Fourteen metal tubes, like short sections of shiny drainpipe, poke upwards from a flat metal table. From the bottom of each, a tangle of wires emerges. It's a kind of radio telescope, though as different from Jodrell Bank as you could ever imagine. 'This is the Very Small Array,' he explains. 'We named it in a kind of tongue-in-cheek way to contrast it with the Americans' extremely expensive and extremely clever Very Large Array in New Mexico.' Small but perfectly formed for its task, the Very Small Array is

ABOVE *The Very Small Array* (foreground), *at an observatory on Tenerife, tunes into the afterglow of the Big Bang – the ultimate act of astronomical violence.*

specially designed to tune into the heat from the Big Bang – heat that reaches the Earth today in the form of radio waves. 'When the Universe was very new, very young, it was very small and very hot. And it's been expanding all the time since then. As it's got bigger the radiation has cooled, and that's why rather than being terribly hot now – hundreds of thousands of degrees or more – it has a temperature of just below three degrees absolute; that's about minus 270 degrees Celsius.'

Despite its cool temperature, this radiation from the Big Bang is the hottest topic in cosmology today. The Very Small Array is competing with over a dozen other telescopes to check out the primeval cosmic fireball. Others have been sited at locations even more exotic than the Canaries, including telescopes in the Antarctic and two that have observed the sky as they hung from balloons.

But the prime location is space. In the early 1990s, a NASA satellite looked far enough out into space, and back in time, to make the first thorough check on the radiation from the Big Bang. The *Cosmic Background Explorer – COBE* for short – looked as far out in space, and as far back in time as any telescope can ever hope to peer. It was staring at gases from the primeval fireball that lie thirteen billion light years away; thirteen billion years back in time.

The Universe was then a third of a million years old, so even *COBE* was not seeing back to the instant of the Big Bang itself. In fact, trying to look further out into space is well-nigh impossible. Beyond the point that *COBE* saw back to, the Cosmos was filled with an opaque fog of superhot gases. These shroud the moment of creation from any telescope that astronomers can yet build.

But that doesn't stop theorists from calculating what happened from the moment of the Big Bang onwards. David Spergel from Princeton University says: 'The radiation was first created only a minute or so after the Big Bang, when the temperature was millions of degrees. It was constantly colliding off electrons, so the Universe was opaque – you couldn't see through it.'

As the Universe expanded, the gaseous fireball cooled off. Some three hundred thousand years after the Big Bang, its temperature fell to three thousand degrees Celsius – cooler than the surface of the Sun, but still far hotter than a steel furnace.

At that point, a profound change took place. Heavy subatomic particles called protons grabbed hold of the free-flying electrons. They hitched together to form atoms of hydrogen. This is the lighter-than-air gas used in early airships like the *Hindenberg* and the *R101*: as their operators found, hydrogen burns fiercely in contact with air. But for cosmologists, the important thing about hydrogen gas is that it's transparent. Light goes straight through it.

In an instant, the Universe cleared: it changed from a hot opaque fog to a clear gas. For a particle of light – a photon – the clearing would have been both abrupt and magical.

'The picture I always have in mind,' says Spergel, 'is when I'm flying in a plane through a cloud layer, where the light is trapped by scattering off of the tiny droplets in the cloud. Suddenly you emerge from the cloud layer, and the air is clear and you can see a great distance.

'Well, if you were a photon of this background radiation,' he continues, 'that would be your experience too. Initially you'd be trapped in this dense fog; and suddenly the electrons and protons would combine to make hydrogen and you would emerge from the fog – and then you'd propagate freely for the next thirteen billion years. You get thirteen billion years of freedom, and then you get absorbed in one of our detectors!'

Instruments like the Very Small Array are indeed hungry for these whispers of heat from the primeval fireball. But in which direction should they peer, to perceive this ultimate in cosmic violence? The answer is – any direction you care to look. The radiation from the Big Bang comes to us from all over the sky.

'That's because the Big Bang didn't take place at a point,' explains Spergel, 'but rather it took place every place in the visible Universe at the same moment. Right where we are today, there was a Big Bang, and the radiation from here has moved off thirteen billion light years in all directions.

'Similarly, the Big Bang took place in regions thirteen billion light years away from us, and the photons from there have been streaming towards us ever since. So this background radiation comes to us from all directions – because the Universe was hot in all directions.'

But it was not equally hot. The object for cosmologists like Saunders and Spergel is to take the temperature of the fireball in different directions in the sky; to check out patches that are slightly hotter and slightly cooler.

'A principal aim of this work,' elaborates Saunders, 'is to understand how structure in the Universe came into being. The Universe today is full of structure, on all scales. Planets orbit stars; stars are clumped together, held by gravity into galaxies; and these are clumped together in clusters of galaxies – and those are about the biggest building blocks we see. So the Universe is concentrated in patches, with pretty much voids in between.'

Suppose the original fireball had been a perfect ball of hot gas – the same absolutely everywhere. Then there would have been nothing to make it curdle into denser patches. 'If you figuratively turned on gravity in an early Universe where the matter distribution was completely smooth, then every bit would be pulled equally by every other bit,' Saunders continues, 'and no clumping would occur, no structures would form. Patently, that's not true.'

But for many years astronomers faced a frustrating impasse. Although the Universe is obviously lumpy today, the first hints of radiation from the Big Bang seemed to show it was a perfect fireball.

The breakthrough came with the *COBE* satellite. In 1992, it discovered a pattern of warmer and cooler patches in the primeval fireball. '*COBE* was picking out fluctuations that were actually produced in the first moments of the Universe,' says Spergel. 'What we're seeing with *COBE* are the fluctuations in density that would grow to form the largest visible structures in the Universe – clusters of galaxies; superclusters of galaxies.'

And, since then, cosmologists around the world have vied with one another to check out the details. While *COBE* was seeing only the large-scale view, the Very Small Array is homing in on much smaller details in the intricate tapestry imprinted in the primeval fireball. 'We're getting results in at the moment,' Saunders enthuses, 'and it's looking very good.'

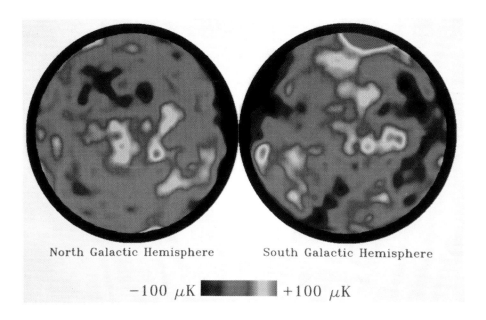

North Galactic Hemisphere   South Galactic Hemisphere

$-100\ \mu\text{K}$ ▮▮▮▮ $+100\ \mu\text{K}$

With the race to probe the primeval fireball, there's an air of secrecy to the experiments. When we ask Saunders what his team has found so far, he replies – only half-jokingly – 'I'm not allowed to tell you!'

But the world of science these days is as much about collaboration as competition. 'Now it's down to the detail,' Saunders continues, 'all the groups doing this kind of work are going to have to look at their data together. Each group uses a different method of observing, and these all have associated errors with them. We have to tease out these problems. So we have to do the work carefully, rather than rush just to be the first.'

ABOVE *The fireball from the Big Bang, as 'seen' by the* COBE *satellite in 1992. Each circle covers half the sky, with hotter regions of the fireball coded red and cooler parts blue.*

But there's one competitor that's creating a sense of urgency. The *MAP* spacecraft is head and shoulders above the others, observing from the clear environment of space. It will be a big improvement on its predecessor, *COBE*.

'Where *COBE* made a measurement,' says Spergel, '*MAP* will make a thousand. If we were thinking about taking a picture of the Earth, then *COBE* got that first picture: it saw oceans and it saw land. What *MAP* will do is now zoom in and see the mountains, the continents, the detailed structure.'

Theorist Spergel has some predictions as to what these 'mountains and continents' in the fireball might look like. But he also privately hopes he won't be entirely correct. Any deviation from our current theories could provide vital clues to the moment the Universe was born.

ABOVE *Investigating the heat of the Big Bang from the chill of Antarctica: a giant balloon will loft the Boomerang telescope (right) to the edge of space.*

'When we get back to the Big Bang,' says Spergel, 'we need to simultaneously use two great theories of twentieth-century physics – quantum theory and general relativity. But at those earliest moments they contradict each other. Now my fantasy is that *MAP* will see something that's not consistent with our expectations, and that might point the way to unify quantum mechanics and gravity. That's certainly not guaranteed, but it would be something wonderful.'

For most cosmologists, all the data returned by *MAP* will be wonderful. It will be a cornucopia of information on the primeval fireball, revealing how the Universe was blossoming three hundred thousand years after its birth. And the fingerprints of warm and cool patches is a crystal ball – properly interpreted, they can reveal not just the past of the Universe, but its future.

In the 1920s, American astronomer Edwin Hubble made the astounding discovery that the Universe is expanding – clusters of galaxies are rushing apart from one another. So it seems our fate is to live in a Cosmos that grows ever bigger, and colder. But there is an alternative. The galaxies are pulling on each other through their gravity, just as Isaac Newton's famous apple was pulled to the Earth. If there's enough matter in the Universe, it could reach a maximum size – and then shrink back on itself, faster and faster, until it disappears in a blazing Big Crunch.

'Some people say the Universe will die in fire,' explains Spergel, 'others think it will die in ice. In the fire version the Universe collapses on itself, it's a very dense Universe; in the ice Universe we just expand for ever.'

To forecast our ultimate fate, astronomers are drawing together the two great themes of the large-scale Universe. On the one hand, we have maps of the cosmic fireball. And on the other, we have the ultimate cosmic zoom from the Anglo-Australian Telescope: the 2dF survey of blazing quasars, distant from us – but not nearly as remote as the primeval fireball.

'With optical telescopes,' says Richard Saunders, 'we can look at large-scale structure roughly at the present time. The microwave background comes from a period when the Universe was a thousandth of its present size. Comparing them will tell you very fundamental things about the physics that's going on.'

In particular, cosmologists now think there's some fundamental physics in the Universe that was hardly suspected a few years ago. Astronomers chasing the most distant of exploding stars, supernovae in remote galaxies, have evidence that the Universe is not just expanding, but accelerating.

If this is right, our path to the icy future will be travelled faster and faster. And what's driving the acceleration, according to the most popular theory, is an invisible force permeating all of space. It's called dark energy.

'The combination of *MAP* and other experiments like the 2dF survey will tell us how much dark energy we have,' says David Spergel. 'If we're very lucky it may tell us something about its properties, too.'

'One of the tests we can do with the 2dF quasar survey,' expands Brian Boyle at the Anglo-Australian Observatory, 'is to check the shape of the quasar clusters. On average, we'd expect them to be spherical.'

To analyse the 3D shape of a quasar cluster is a complex business. You can see its extent on the sky quite easily; to work out the third dimension, its depth, you have to make some educated guesses about the way the Universe is expanding. And the only way you can make the quasar clusters come out spherical – on average – is to add a large dash of dark energy to empty space.

According to Boyle, 'What we've found out is quite consistent with what we're learning from things like the microwave background and the supernova measurements. It's all hanging together very well.'

At the start of the third millennium, astronomers have achieved what even a generation ago would have seemed impossible. Giant telescopes have charted the Universe, taking in explosions of more and more awesome proportions, from supernovae to hypernovae and quasars. Miniature radio telescopes, on the ground and in space, have extended our cosmic zoom to allow us an intimate acquaintance with the ultimate act of violence, the fireball from the Big Bang.

Along the way, astronomers have charted an invisible force in the Universe that has power over all others: the dark energy that is driving us to our inexorable fate, an icy and empty Cosmos.

With these all-embracing discoveries, have we now reached the point where astronomers understand the Universe fully – where they can mothball their telescopes in the knowledge that nothing remains to be discovered?

'We are making giant steps forward,' demurs cosmic cartographer Brian Boyle, 'but we shouldn't kid ourselves. We're doing great work; we're asking fundamental questions. I think it's wonderful that humankind should even have the temerity to ask these sorts of questions, and also have the maturity to be able to think of some of the answers.

'But are we ever going to find an ultimate answer? No, not in my lifetime. But that's half the fun of it. I'm enjoying the journey more than the destination – because I know I'll never reach that destination.'

In an infinite

Towards

chapter eight

Universe, there must be a planet where Elvis lives!

Infinity

In medieval times, it was an easy question to answer. You wanted to know where's the edge of the Universe, and what lies beyond? Well, the Scholastics had the answer. Merging the ancient Greek theories of the Universe with the Bible, they had the Cosmos mapped out.

At the centre is the cesspit of the Universe: that's the Earth. Inside the Earth, it's literally downhill all the way to the fires of Hell. Moving upwards, you enter more and more sublime realms. The Moon is nearer perfection than the Earth, but it still has blemishes on its silvery face. Further up, we find the planets. Though they may be faultless points of light, their motions are not quite regular. The planets make up for that imperfection, though, by singing out the sublime 'music of the spheres' as angels push them around the sky.

Finally, outside the crystalline spheres of the planets, you reach the great black sphere that carries the shining stars. This is perfection. The smooth nightly motion of the star-sphere derives its power from the Prime Mover, God himself. This power permeates down to the angels on the planetary beat, and in debased form descends to the Earth itself.

The great fourteenth-century Italian poet Dante acknowledged this natural order in his *Divine Comedy*, a journey from the depths of Hell up to the highest realms: 'The glory of Him who moveth everything, doth penetrate the Universe.'

ABOVE *An ancient traveller peers beyond the 'edge of the Universe' in this Victorian woodcut. Most medieval scholars, though, would expect Heaven to lie behind the dome of stars.*

For Dante and his contemporaries, the solid sphere of the stars was indeed the edge of the Universe. Beyond it we leave our Cosmos altogether, and move into the otherworldliness of God's eternal paradise. 'Within that Heaven which most his light receives was I,' Dante recounts, 'and things beheld which to repeat, nor knows, nor can, who from above descends.'

In the twenty-first century, our growing knowledge has put astronomers in the position of having to employ language a trifle less poetic.

'It's important to understand that – at least in the simple conventional cosmological models – there isn't an edge at all,' explains Paul Davies, a British physicist who's worked for many years in Australia. 'The Big Bang was the appearance of both space and time from nothing, and at all times matter was distributed more or less uniformly across all of space.'

'There should be no edge to the Universe,' concurs Janna Levin, an American physicist researching in Cambridge, England, 'because it would make the laws of physics look like things we can't understand. There's no way of living outside of space-time, so the concept doesn't really make sense.'

Like Alice, though, modern cosmologists live in something of a Looking-Glass world. Everything is not always as it seems, as Davies cautions: 'Well – the definition of the Universe is a little bit ambiguous. We have to distinguish between the Universe we see, and the Universe of everything that there is.'

To the frustration of scientist and layman alike, the Universe we can see is only a tiny portion of 'the Universe of everything that there is'. The culprit is light. When we look into the depths of the Cosmos, we see everything as it was in the past.

When the Sun shines into your eyes, you are seeing light that left it eight minutes ago. On a clear night, look at the bright star Rigel in the famous constellation of Orion, the hunter, and your eyeballs are receiving light that left this giant star 800 years ago, before Dante set quill to parchment. The further away a star or galaxy lies, the further back in time we see it. Astronomers with giant telescopes now routinely survey galaxies as they were before the Earth was born.

For many astronomers, this natural time machine provides a fantastic ride back into the past. But it's a source of sheer frustration for those who are on the trail of the edge of the Universe.

'The distance we can probe depends on the instrument we're using,' says Davies. 'If we're talking about optical telescopes, then with the Hubble Space Telescope we're getting to a point as far back as the formation of the galaxies. With microwave instruments, we can look back to a period which was about three hundred thousand years after the Big Bang; we could even imagine some sort of telescope that could take us back to one second after the Big Bang, and then we'd be looking back practically as far as we possibly could.'

How far back in time is that, compared to the present day? 'Well, the answer to that question depends on a number of things that we don't know precisely,' Davies continues. 'But it was something probably between twelve and fifteen billion years ago.'

So, wherever we look out into space, we are limited in our view. However powerful our telescopes, we can never hope to see anything before the Big Bang. Even using Davies's biggest estimate, we can never peer more than fifteen billion light years from home. Within a giant sphere stretching fifteen billion light years from the Milky Way lies everything we can ever hope to see. It's 'the observable Universe'.

The size of the observable Universe is staggering in everyday terms – almost a million million million million miles. Yet, if the entire Universe has no edge, then our observable Universe must be only a trifling part of the whole. And astronomers are unanimous in believing that edge of the region we can see is not the edge of the entire Universe – space does indeed carry on beyond.

Davies reiterates, 'We can't see a distance greater than light has travelled since the beginning of the Universe, and so there is an effective boundary to what we can

ABOVE *This is the furthest view we have into the depths of space – the Hubble Deep Field. Some of these galaxies lie over twelve billion light years away, near the edge of the 'observable Universe.'*

see. But it's not a real physical edge, it's just like a horizon. If you go up a mountain and look out to sea, it looks like the world has an edge there. But of course we know it doesn't.'

From a mountain-top observatory in Tenerife, Cambridge astronomer Richard Saunders has seen as far back in time as anyone. His Very Small Array is picking up whispers of radiation from the Big Bang – close to the horizon of the observable Universe.

And, from here, Saunders looks out to a more familiar horizon – the 'edge of the world' that Columbus sailed over on his pioneering expedition to the New World. 'We can only have direct ability to determine things that we can measure,' he says, 'and so necessarily we can only see out to a distance that corresponds to the time it takes light rays to travel from the start of the Universe to us. But there's no reason to suppose that if we were transported elsewhere in the Universe, things would look much different.'

And so Columbus found when he got to the Canary Islands' horizon. Though his ships disappeared from the sight of those watching from Tenerife, Columbus didn't fall over the edge of the world. All he found was much more of the same – sea, sea and more sea.

If we could transport ourselves *instantly* to the edge of the observable Universe, we'd find ourselves in much the same boat as Columbus. In the billions of years since the Big Bang, the cosmic fireball that Saunders is observing has had time to grow up.

ABOVE *When a sailing ship disappears over the horizon, it still exists although we can't see it. Similarly, many unknown galaxies may lie beyond the 'horizon' of our observable Universe.*

Its gases have congealed into galaxies, stars and planets. Transported here in an instant, we would find ourselves surrounded by a Universe that looks comfortingly like our home region around the Milky Way.

If there's an alien race out here, they'd see the Milky Way off on the edge of their own observable universe. There'd be one difference from Columbus's situation, though – they'd be looking back in time, so they would see our Galaxy as it was in its earliest youth. In this Looking-Glass Universe we inhabit, we see everyone else as they were long ago, and they see our past.

When these aliens turned around to look in the opposite direction to the young Milky Way, they'd see more galaxies stretching off into space. These galaxies are invisible from Earth, even with the most powerful telescope: they lie beyond our horizon. Paul Davies explains the problem clearly with a simple analogy. 'If you imagine three ships in line astern, out on the ocean, then the lookout on the middle ship might just be able to see through a telescope the ship ahead and the ship behind. But these leading and trailing ships couldn't see each other, because they're twice as far apart, and they're over each other's horizon.'

For Columbus's three ships, keeping within the same horizon was a practical necessity. By the time they'd crossed the Atlantic, they'd travelled an immense distance beyond the horizon of the Canary Islands – way beyond the realm that anyone in the Old World could see. Here they were greeted by inhabitants of the Bahamas, who lived in a world bounded by their own horizon. However hard they gazed eastwards, they could never see Europe.

Similarly, there could be 'new worlds' way beyond the limits of our cosmic horizon, for ever beyond our ken. 'Though we can't ever be one hundred per cent sure,' says Davies, 'presumably there will be galaxies beyond our visible horizon that could well have inhabitants who would not be able to see our Galaxy. And we can't see theirs. Each of us is surrounded by a finite horizon region.'

For theorists like Davies, it's incredibly liberating to be able to stand back and – in his mind's eye – to imagine a Universe beyond the horizons that are imposed by the dawdling pace of light. It's like viewing ships on an ocean from a high-flying plane. Instead of being constrained to see only a small circle of ocean within our own horizon, we can now survey the entire ocean.

This immensely much broader canvas takes us further along the quest to check out whether there is an 'edge of the Universe'. What would happen if we used our bird's-eye view to survey the Universe further and further out, zillions of times further than the horizon of our observable Universe? Or is this quest as meaningless as the idea that Columbus would fall off the edge of the Earth?

'No, you don't fall off the edge of the Universe,' says Janna Levin. 'Imagine a sphere like the Earth – it doesn't have an edge, it's a completely connected surface and

ABOVE *In this mega-bird's eye view, the observable Universe around the Milky Way (blue) is bounded by our horizon (blue circle). We can't see the red galaxy. Conversely, its inhabitants have a view bounded by their horizon (red circle) and can't see us.*

it's completely self-contained. The surface of the Earth is finite: it doesn't go on for ever, but nor is there an edge.'

Warming to her theme, she continues: 'There's no point in saying that England is the edge of the Earth – right – any more than saying that France is the edge of the Earth, 'cause there really is no edge, it's all continuous and connected. And so that's something that could happen in the Universe: it has no edge, but wraps back on itself.'

Ferdinand Magellan had faith that the Earth would wrap back on itself when he set out in 1519 on the first circumnavigation of the globe. Though Magellan himself died on the way, his crew returned safely to Spain, after sailing more or less a straight course through the world's major oceans. 'And so with the Universe,' Levin adds. 'In theory, if we headed out in a spaceship, carried on in a straight line without stopping or turning, we would eventually come back to where we started.'

Paul Davies has a different view. 'It could be that space is infinite. It doesn't have an edge, but just goes on for ever.'

Infinity is a mind-boggling concept. Our finite minds can't really comprehend the idea. But, then, we never really have to. We've been brought up on a small world, and even the vast reaches of the observable Universe have a limited extent.

In an infinite Universe, space just goes on and on. However far – and fast – you travel, there is never an edge. You'll pass galaxy after galaxy along the way, literally without number. Stars after stars; planets after planets.

And there are more mind-numbing thoughts to come. A planet is put together from a certain number of atoms – albeit a huge number. And there's only a certain number of ways of putting these atoms together to make up planets. Now, the number of possible planets may be far bigger than all the grains of sand on all the beaches on Earth, but it's still a finite number.

If the Universe is literally infinite in size, and you had unlimited time, you could find *all* these different possible planets. Then you'd start finding doubles of the planets you already know . . .

'If the Universe is infinite in spatial extent and uniform,' smiles Davies, 'then it is absolutely certain that there will exist other beings identical to you and me. It is one hundred per cent certain that the entire inhabitants of the Earth will be repeated, with duplicate Paul Davieses and Nigel Henbests.'

Perish the thought! But Davies does help to put it into context. 'The distance between us and this duplicate Earth is stupendously large, far larger than the size of the observable Universe,' he adds, 'but in an infinite Universe you've got plenty of numbers to play with.'

And, in addition to the exact duplicate worlds, you'll have other planets that are just a little different from our actual Earth. In an infinite Universe, there must be a planet where Elvis lives!

Despite the tabloid headlines involved in this claim, Davies downplays the consequences of the infinite Universe. 'This is really a mathematical curiosity,' he cautions, 'rather than anything of terribly important physical significance. Because when you're dealing with infinity you can come up with all sorts of weird and wonderful results.'

Janna Levin certainly has her doubts about an infinite Universe. 'I'm not sure if there's a problem mathematically so much as physically. I think it's absurd!

'If we really believe the Universe started in a Big Bang,' she continues, 'then it becomes really untenable to imagine it came out of nothing and was instantly infinite – that from nothing came an infinite amount of space, matter and energy.'

Despite the consensus among her colleagues, Levin puts her money on a Universe that doesn't go on for ever. In her view, it bends back on itself like the Earth's surface – finite in extent, but without any edge.

Can we ever hope to tell which theory is right? Perhaps not. But cosmologists see this question as a major challenge. They're hot on the trail, squeezing every last bit of information they can out of the Universe that we can see. Their spiritual guide on this quest is Albert Einstein – the obscure German patents clerk who, a century ago, drew up a staggering overview of the entire Universe.

At the time, astronomers didn't even know about other galaxies. Many of them thought that our Milky Way was the entire Universe. Yet Einstein's thoughts ranged far beyond what telescopes could see. In his mind's eye, he could see that space might be curved.

The key to the shape of space, Einstein proved, was the amount of matter and energy around. You can see this most clearly when a lot of matter is concentrated in one place. In the middle of a star that's gone supernova, there's a dense collapsed core that distorts the space around it. The core – a neutron star, perhaps a magnetar – makes a deep dent in space, like a heavy ball sitting on a sheet of rubber.

If the core shrinks more, concentrating its matter into a smaller and smaller ball, it will sink deeper and deeper into the stretchy fabric of space. Eventually, space gives up the unequal struggle. The dent in space becomes a bottomless well. The dying star's core has become a black hole.

ABOVE *Albert Einstein took on the challenge of gravity when he formulated his theory of relativity: to this day, it is the basis for understanding the Universe.*

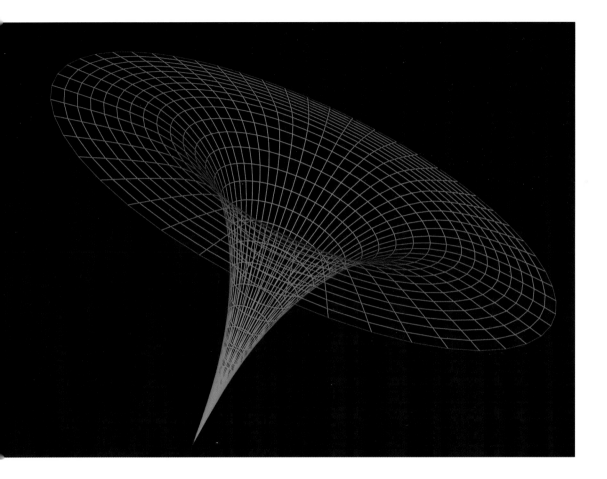

Because the black hole is a fathomless pit in space itself, anything that falls in cannot get out. And that includes a beam of light. Shine a torch onto a black hole and you won't illuminate it. No light is reflected back to you; the most powerful beams simply follow the curve of space and disappear down the well.

Not content with that intellectual tour de force, Einstein had the chutzpah to apply his theory to the opposite extreme – from tiny collapsed stars to the whole vast extent of the Universe.

Again, matter has the whip-hand, bending space itself into curves. But over the vast reaches of space, matter is spread out thinly. Instead of creating a vertiginously steep hole, it gives the whole of space a gentle curvature.

Heavy stuff indeed! How can ordinary mortals imagine what Einstein proposes: curvature of our three-dimensional space into other dimensions? The short answer is that's literally impossible for the human mind to visualize. In fact, Janna Levin cautions that we shouldn't even try.

ABOVE *A black hole is best understood as a funnel in space, according to Einstein. Fall down its steep sides, and you'll never escape.*

'We can't ever leave space-time and look at it from the outside,' she says, 'there's no meaning to that statement. Where would we be standing, and how long would we be there? There's no meaning to those questions unless there's a space and a time to stand in. So we can't stand off of the Universe and look at it from a distance, and say, "It looks this shape." '

With all deference to Levin, the Looking-Glass world of modern cosmology prompts us to follow the White Queen's advice when Alice insists that one can't believe impossible things. 'I dare say you haven't had much practice,' says the Queen. 'When I was your age, I always did it for half an hour a day. Why, sometimes I've believed as many as six impossible things before breakfast.' So, whether you've indulged in breakfast yet or not, let's try just one impossible cosmological thing – to see our Universe from outside.

To set the scene, imagine a universe that has only two dimensions. Instead of our 3D universe, where you can go left–right, forward–backward, and up–down, the 2D universe is like a thin sheet of paper. The flattened inhabitants live in the plane of the paper. They can only move left–right and forward–backward; they have no idea of the up–down dimension.

Standing outside their universe, we can view it from above and – godlike – survey the restricted life of the flat creatures. And we can see the shape of their universe. It may be flat, like a sheet of paper. Or it could curve round, like the skin of a balloon. And these aren't the only possibilities. Albert Einstein suggested that space could also be shaped like a surface of a saddle: instead of curving back on itself, it flares away in all directions.

While we're in Looking-Glass land, here's another odd fact. If the 2D universe is totally empty – no inhabitants; no planets, stars or galaxies; no matter whatsoever – then it will twist itself in the saddle shape. It's a saddle that goes for ever, out to infinity.

Sprinkle some matter into the 2D universe and, according to Einstein, the curves begin to smooth out. Add some more matter, and you can make this universe go flat – an infinite sheet of flat paper. Still more matter, and the 2D universe curls around into the shape of the balloon skin, connecting itself into a ball: it's a universe that has a limited size, but doesn't have an edge.

Sounds familiar? Indeed, these different possibilities for the 2D universe are exactly the same as those the astronomers postulate for our own Universe: infinite in extent, or wrapped around to connect up with itself a long, long way off.

So now for the real impossible task: instead of a 2D universe curving up and down, imagine our 3D universe curving into the fourth dimension. Can't do it? Well, even Einstein couldn't quite achieve that. But the maths is much the same.

And it gives us some clues as to how to check the shape of our Universe. The key is the amount of matter about. If there's too little, our Cosmos will flare away in an

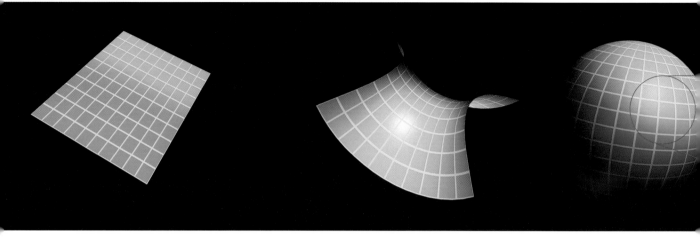

otherworldly saddle. If there's too much matter, the Universe will curve around like the surface of a multidimensional balloon. Or we could live in a Goldilocks Universe: not too little matter, not too much matter – just the right amount of matter to make the Universe flat.

Leading the investigation is the Anglo-Australian Telescope, towering above the kangaroos and gum trees of New South Wales. Here, astronomers have been compiling a huge census of galaxies. This cosmic survey has allowed them to weigh up the matter in the Universe.

'What we're doing,' explains the observatory's director Brian Boyle, 'is to look at the shape of clusters of galaxies. They can look slightly squashed, because of the effects of gravity making things fall into them. Once you work out how much gravitational attraction is there, you can work out how much mass is there.'

Other astronomers are weighing up these clusters of galaxies with speeding beams of light. Gravity bends and focuses the light winging its way from distant quasars. Like your reflection in a fairground Hall of Mirrors, the image of the quasar becomes distorted. Instead of a sharp point, it appears as a curved line of light, like a portion of a spider's web.

Whichever method you use, the answer is the same. 'The amount of matter we measure,' concludes Boyle, 'is thirty to thirty-five per cent of the amount required to close the Universe.'

So we live in a lightweight cosmos. According to these cosmological weigh-ins, there's not nearly enough matter to pull the Universe from a flaring saddle shape to a flat Cosmos – to close the Universe, as Boyle puts it.

But matter may not be the only thing that matters. In Einstein's world, there's more than one way to bend space. The ingredient that most astronomers have ignored

ABOVE (LEFT TO RIGHT) *A 2D universe can come in various shapes. It can be flat (left); curved like a saddle (middle) or curved like the surface of a ball (right). The real 3D Universe could similarly curve into a different – and unimaginable – dimension.*

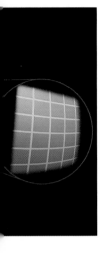

until now is energy – 'dark energy' that might be spread out, invisibly, throughout the Cosmos. And dark energy has a strange property. It pushes outwards on everything, making the Universe expand faster and faster.

In the late 1990s, dark energy revealed its whip-hand over the Cosmos in the most dramatic way. Rival teams of astronomers were investigating supernovae in distant galaxies, and discovered they were racing away from us at ridiculous speeds. The accelerating Universe hit the headlines. And, by and large, it astonished the scientific world.

One scientist who wasn't surprised was Paul Davies. 'I rather stuck my neck out in the early 1980s when I wrote a book called *About Time*,' he says, 'which basically came out in favour of this accelerating force – rather against the scientific trend at the time. And I was very pleased that in the fullness of time, the observations seemed to bear this out.

'I've always had a soft spot for this accelerating issue,' Davies continues. It started with his deep respect for Einstein – a respect that went one step beyond Einstein's faith in his own equations. These equations describe gravity and the shape of space. 'In 1917, Einstein put a repulsive force into his equations – sort of by hand,' says Davies, 'because he recognized a problem. How is it that all the galaxies can just

ABOVE *Ghostly curved galaxies here are cosmic mirages, their light bent by the gravity of the big galaxies in the foreground. Such views allow astronomers to 'weigh up' the matter in the Universe.*

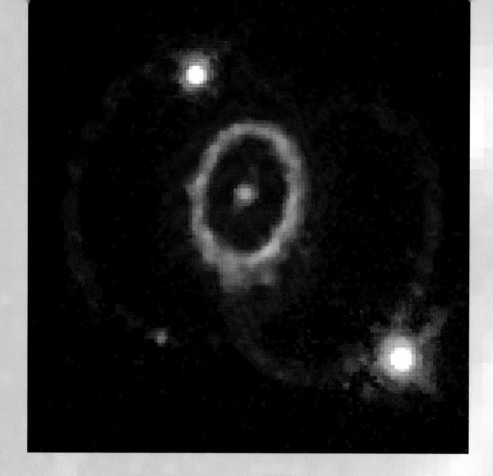

hang out there without all collapsing together, if there's only one force in town – and that's the attractive force of gravity?'

Today, it's a problem no more. We know the Universe is rushing apart from the Big Bang. Pure momentum can keep the galaxies from collapsing together, without having to invoke a repulsive force. When astronomers discovered the expanding Universe, in the 1920s, Einstein felt he didn't need the repulsive force any more. So he took it out of his equations – and called it his greatest blunder.

A few decades later, other physicists took a different look at the nature of space – a microscopic view. 'Quantum mechanics suggests that even empty space is not totally empty,' explains Davies, 'it's a sort of feeding frenzy of subatomic particles coming into existence and disappearing again – flitting in and out of existence the whole time.'

These particles carry some energy with them; and a straightforward calculation ends up with the embarrassing suggestion that they would fill space with an infinite amount of dark energy. Its infinitely powerful outward push would have blown the rest of the Universe out of sight an instant after the Big Bang. As it is, the Cosmos has clearly survived its inbuilt acceleration for billions of years.

'And so you might say – well, Einstein wants this to be zero; quantum physics suggests it's infinity; so why not take an average and say it's somewhere in between!' Davies laughs.

ABOVE *Remains of nearby Supernova 1987A form interwoven rings in the sky. More distant supernovae have proved the Universe is accelerating: expanding faster and faster.*

OPPOSITE *Subatomic particles perform intricate loops in a physicist's lab on Earth. On the largest scale, they may make up the 'dark energy' that powers the accelerating Universe.*

And somewhere in between it seems to be – but how much in between? The supernova hunters have a rough idea, but for the definitive answer it's back to measuring the shape of the Universe. Because dark energy – just like matter – will help to flatten out a saddle-shaped Universe.

To measure the curves of the Universe, we need a vast tape measure. And on this great ruler, we need to have regular ticks, marked not in inches or centimetres but in millions of light years. That's a pretty tall order for any cosmic tailor – but it just so happens that Nature has provided astronomers with what they need. It's a giant measuring grid that covers the whole sky, marked out with a roughly regular pattern of spots.

This is the last whimper of the Big Bang – radiation from the original cosmic fireball, now diluted down to a faint glow of radio waves spread across the sky. It's textured with warmer and cooler patches.

'What we've realized is that we can look for a pattern in the hot and cold spots in the sky,' enthuses cosmologist Janna Levin, who's been sufficiently inspired to write a whole book on the topic. 'I've called it *How the Universe Got Its Spots* – and I guess you in England will be more aware of the allusion in the title than people back in the States.' While Rudyard Kipling's leopard was given his five-fold spots late in life, as fingerprints from an Ethiopian, the Universe has been spotty since birth. And we can put them down to the infantile screams of the baby Cosmos.

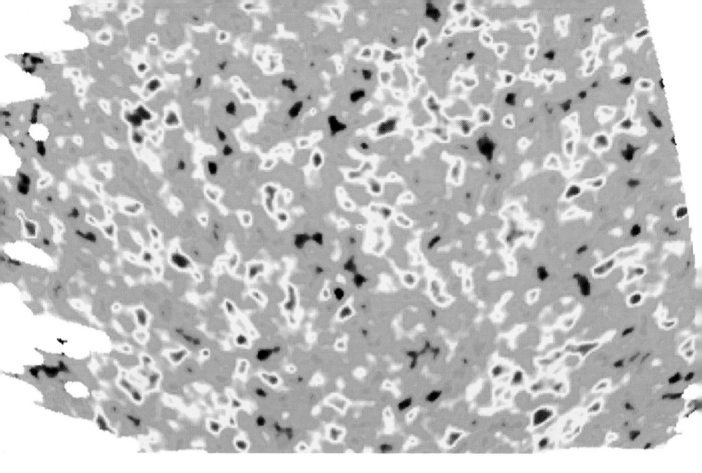

Just after the Big Bang, the seething fireball was filled with unimaginable noise. Sound waves ricocheted back and forth. The whole Universe resonated, like a set of giant organ pipes. In a concert hall, music wafts to our ears as series of waves in the air: the noise of the Big Bang made correspondingly vast waves in the hot gas. The pattern of these vast sound waves was etched into the fireball, and astronomers observe them today as warm and cool spots in the sky.

'What I'm looking at,' continues Levin, 'is partly how this pattern could be imprinted in the Universe. And also how these hot and cold spots can tell us about the underlying geometry of space.'

So, for Levin, the spots have a very practical function: they are the markers on the cosmic ruler that we need to measure the shape of the Universe. 'The wavelength of the fluctuations,' she explains, 'is very tightly linked to the curvature of space. So by looking at the size of the spots – what we call the Doppler peaks – we can determine whether the Universe is curved or flat.'

Around the world, and from satellites in space, dedicated teams of astronomers have been scrutinizing the spots in the Universe. The roll-call began in 1992, when NASA's *COBE* satellite picked out the biggest spots. Telescopes in Cambridge, Antarctica and Canada, as well as the Very Small Array in the Canary Islands, have been helping to fill in the details. And a new space mission, *MAP*, is poised to take the finest snaps of the Universe in its infancy.

ABOVE *The spotty Universe: this image from a balloon-borne telescope called Boomerang reveals patterns in the cosmic fireball that created our Cosmos.*

In the meantime, the best views have come from two telescopes that have ascended to the verge of space. The Boomerang and Maxima telescopes both flew on balloons, and discovered an intriguing pattern in the cosmic spots.

'Boomerang and Maxima and those other ground-based and balloon-borne experiments have done really well so far in terms of measuring the Doppler peaks,' says Levin. 'And we can now draw most of the really huge conclusions.'

Leaving aside all the other exciting cosmology, what's the answer to the shape of the Universe? 'They show that the Universe extends beyond what we can observe, and that it's probably very nearly flat.'

A Universe that contains only matter, though, is only one-third of the way to being flat. Something else must be helping to bend the cosmos. And according to our best theories – the balance between Einstein and quantum physics – there's only one other possible ingredient in the cosmic mix. And that's dark energy.

Cosmic weigher-in Brian Boyle summarizes: 'What we're learning from the microwave background – along with our 2dF survey and the supernova constraints – is that the amount of matter is about thirty to thirty-five per cent of the Universe; and the remainder – sixty-five to seventy per cent – is in dark energy. It's all hanging together very well.'

Like a pair of heavyweight twins, matter and dark energy pull together to flatten out the fabric of space. From a philosophical viewpoint, a flat Universe certainly has a kind of simple beauty. But is there a good scientific reason why it should be flat, rather than saddle-shaped or curved in on itself?

'Well, there's a strong theoretical prejudice that the Universe should be flat,' comments Janna Levin. This prejudice comes from a popular theory for the early Universe. According to the theory of inflation, the Universe, in the first moments after the Big Bang, suddenly got vastly bigger. Like Topsy, it 'just growed and growed'. The reasons for inflation are hidden deep within quantum physics, but it makes one remarkable prediction: the Universe should appear flat.

Levin brings inflation theory down to Earth for us. 'It's like saying, when inflation started we might have been standing on a basketball,' she explains, 'and we can see that that's curved. And inflation made it so big it's like we're standing on the Earth. It's so big we mistake our local region for being flat.'

If inflation is right – and there isn't a better theory on offer – then we can begin to speculate about what may lie outside our observable Universe.

'Inflation really did push things well beyond our observable Universe,' continues Levin. 'What we did was to start with something which was lumpy and erratic, and blow it up so big that we're insensitive to the bumps. Now, our observable Universe may be almost flat, but the Universe beyond the region we can see may still be incredibly lumpy.'

And there's no reason why these lumps and bumps should be anything like our region of the whole wide Universe. When Paul Davies previously mused on the existence of an infinite tribe of Paul Davieses in an infinite Universe, he carefully put in the caveat that he was talking about a Universe that was 'uniform'. In other words, we'd find much the same kind of atoms everywhere, under the sway of the same kind of electric and gravitational forces that we know at home.

'But there's no reason why these strengths should be God-given,' he elaborates. 'We might find that if we were to travel to a very, very distant part of the Universe that the strong nuclear force – for example – was a bit stronger or weaker. In fact, some people think that almost all what we think of as the "inbuilt laws of physics" are up for grabs.'

If we could magically travel to one of these other regions, we could be in for a tough time. 'It's generally acknowledged by people who play these games,' says Davies, 'that the kind of set-up we have in the Universe we observe is, to coin a phrase, bio-friendly.'

Our local cosmic arena is a place that makes life easy. There are stars emitting light and warmth; planets where we can live; atoms and molecules in our bodies that do just the right things for us to eat, grow and procreate to keep the flame of life alight for aeons.

But in reality, the bio-friendly Universe is traversing a tightrope. Look down to the basics of our Universe, and we discover a supreme cosmic balancing act. Forces of nature counterbalance each other, to make stable atoms and stable stars.

'The existence of life depends rather sensitively on the actual arrangement of the laws of physics and the numbers that enter into them,' Davies continues. 'If we could play God and twiddle the knobs – make electromagnetism a bit stronger or gravity a bit weaker, something of that sort – then there probably wouldn't be stable stars like the Sun and the distribution of elements would be quite different. So there almost certainly wouldn't be life.'

Our own part of the Universe is quite safe, as its laws were indelibly laid down moments after the Big Bang, at the period of inflation. But in far-off regions of space, inflation could have taken a different turn.

We could imagine the entire Universe as an ice cream, made of an infinite number of scoops of different flavours. Our observable Universe is only a tiny crystal embedded in the vanilla lump. The whole vanilla portion has the same kind of physics as us, although the rest of it is too far away for us to see. But – much further still – our vanilla scoop of Universe nestles up to other regions: chocolate, raspberry, rum-and-raisin scoops. Here we'd find physics with an entirely different flavour.

And Davies speculates that there could be 'more drastic' changes. 'For instance, there could be four dimensions of space, instead of three. In that case, you probably

OPPOSITE *An intricate balance in Nature keeps order throughout the Universe, from the giant whorls of a spiral galaxy down to the cells that make up our bodies.*

wouldn't have any stable orbits for planets. And it strains one's imagination enormously – what it would be like to look across a boundary where on the far side there were four space dimensions. It's very hard even to imagine the sort of images we'd get, and how we would interpret them.'

And we may not need to travel that far into the Looking-Glass universe before we start seeing strange images. Any picture of deep space shows us countless billions of different galaxies. But Janna Levin wonders if these really are *different* star-cities. Perhaps there are only a small number of galaxies in the observable Universe, but we get to see them over and over again – like seeing a vast number of images of yourself when you stand between mirrors on opposite walls of a bathroom.

The real Universe doesn't have mirrors, but there could be twists and turns in the fabric of space that focus light in new ways. A black hole is the ultimate twist – a steep funnel that bends a beam of light so steeply that it must travel down the slope and disappear. But space could also bend round much more gradually. It could loop back on itself, like interconnected holes in a Swiss cheese. A beam of light from the Milky Way could follow the warp, and emerge in another part of the Universe – possibly travelling in the opposite direction.

'Very few people think – probably myself included – that we'll ever actually be able to see something like that,' says Janna Levin. 'It's more of a theoretical possibility – but like most theoretical possibilities it may fit into a larger puzzle some day.'

Suppose we set out in a fast starship, and got caught up in one of these curves in space? 'Well, the simplest case,' Levin declares, 'is if we fly in a straight line and come back where we started. But the Universe might not be so simple: we might fly in a straight line and come back upside-down, or come back left-handed instead of right-handed. You might come in upside-down from the left-hand side.'

As with a starship, so with the light from our Milky Way Galaxy. 'It could go through an intricate series of ins and outs around this space and then come back to where it started,' continues Levin. 'So when we look out into space and think we're seeing a different galaxy, it could be that we're really seeing ourselves.'

ABOVE *According to one audacious theory, this apparently distant galaxy may actually be our own Milky Way, seen through a distorting twist in space.*

So a lot of the galaxies we see in the distance could be just copies of our own Milky Way. But it wouldn't be easy to recognize them. Light would take billions of years to travel through the tortuous curves of space, so we'd be seeing the Milky Way as it was long ago: each image we see would be our Galaxy at a different stage of its youth.

Instead of trying to identify individual galaxies, Levin is instead checking out the pattern of spots on the Universe's hide. 'Even if we can't tell if any other galaxy is really our Galaxy,' she explains, 'a multiply-connected Universe should lead to patterns in these spots – things like circles in the sky and other geometrical patterns.'

Levin admits that there's only an outside chance that space is really connected in such complex curves. But she's looking forward to new images of the spotty Universe from the recently launched *MAP* spacecraft. 'When we get *MAP* – and the future Planck mission – we'll be able to do a more detailed assessment.'

And if the curves in space are more extreme, perhaps they could lead to the ultimate in cosmological speculation: other universes, completely detached from our own.

'So far what I've been talking about is a situation where we have one spatial region, all connected together, even if it's chopped up into domains where the physics might be different,' says Paul Davies. 'Now there are other variants where this connection doesn't exist. And one of the easiest to envisage is the "baby universe" model.'

Davies goes back to the analogy of the 2D universe. Instead of a flat piece of paper, though, he wants something more flexible – say, a thin sheet of rubber. 'We could imagine a huge deformation of space, like the rubber sheet having a protuberance erupting from it, almost like a balloon with the neck of the balloon joining it to the sheet.' In the right circumstances, the balloon might inflate of its own accord, growing bigger and bigger. Only its neck connects it to the rest of space, like a narrow umbilical cord. And then that little umbilical cord would pinch off.

At one end, the mother universe reverts to its flat shape. 'At the other end, there's what would seem like a new universe expanding like crazy in its own version of the Big Bang,' says Davies. 'This is one way a universe can give birth to another. And – by extension – this can go on an infinite number of times.'

So universes beget universes. It's a strange procreation. A baby universe and its mother are totally cut off at the moment the umbilical cord snaps. Every mother universe loses its infant at birth. Every new universe is effectively born an orphan.

And that prompts the question of where our Universe fits in. There's no reason why it should occupy a special position in the family tree of universes. Most likely, our Big Bang was an eruption from the distorted space of a pre-existing universe.

'And that raises an interesting question – did the Universe have a beginning?' concludes Davies. 'If by the Universe you mean the volume of space to which we can have access by travelling as far as you physically can, then that Universe did have a

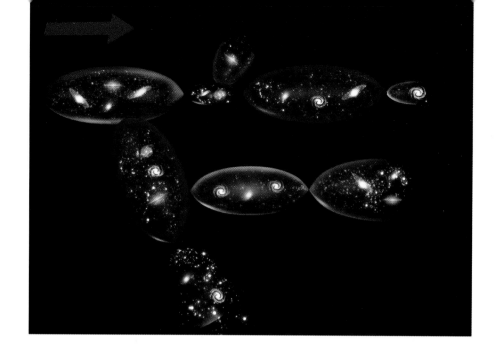

beginning – the Big Bang. But if you take the vast assemblage of all these bubbles and sheets and things, then there is no beginning.'

If we could trace the genealogy of our Universe backwards, through all our forebears and distant patriarchs, we'd never come to a cosmic Eve.

And so we can see a far greater vision than even Dante's revelation of Heaven. We live in an observable Universe that's vast by any everyday measure, but only a smidgeon in terms of the whole Universe beyond. This whole region is undoubtedly *our* Universe: we could reach any part of it by fast spacecraft, given enough time. But when we travel far enough, we're likely to find regions so savage that even Hell seems comfortable by comparison.

Journey further still, and we might travel right round the Universe and back home – to find our Milky Way in the far, far future. Or, more likely, the Universe could be infinite. Carry on trekking, and we'd eventually come across what looks like our Galaxy – but in reality is merely an identical copy. Here, atoms have chanced to come together in the same way to make a replica Milky Way, complete with duplicates of all of us.

But our infinite Universe could be just one bubble in the greater scheme. Other universes fill an infinite number of other dimensions of space. They come and go over a boundless period of time, from the endless past to the never-ending future.

These are awesome concepts. But are they real? In centuries to come, will people view the idea of the infinite Universe and baby universes in the same way that we smile at Dante's vision of the celestial spheres?

'I would love to dream, I'd love to imagine what's going on in other universes – but as a very limited observational astronomer I'm tied to the observations,' says Brian Boyle. 'But I think it's very unimaginative to say, "No, I don't believe there are alternative universes." Because of inflation, I think it's entirely likely that there's lots of

ABOVE *The ultimate view. Our whole Universe* (centre right) *is just one of a series of universes that have grown and procreated since time immemorial...*

OPPOSITE *Only a few centuries ago, no-one could conceive of a giant glowing gas cloud in space like the Orion Nebula. How will our understanding of the Universe mature in centuries to come?*

other universes out there. That's part of our generational chauvinism that I don't think will ever be proved wrong.'

'Given what we understand, I think it's completely plausible that there are inflationary bubbles that branch off,' adds Janna Levin. 'They are separate universes, even if they are roots off the same tree.'

But Levin cautions about thinking of these bubbles too literally. 'As for visualizing them moving off into extra dimensions, that's really just a visual tool. It doesn't have to happen that way; it's just one way for our brains to visualize it.'

And the human brain – despite its awesome mental powers – may be the limiting factor in our frustrating attempts to comprehend the paradoxes we find at the ends of the Universe.

Our current brains are not built to feel infinity, nor to sense other dimensions. Just as dogs could never have much comprehension of galaxies, nor cats of nuclear physics, perhaps our minds just couldn't cope with the ultimate answers, even if they were presented to us.

So our ultimate understanding of the Cosmos may have to await an evolutionary step here on Earth. To comprehend the final frontiers of our extreme Universe, we may need to evolve extreme minds.

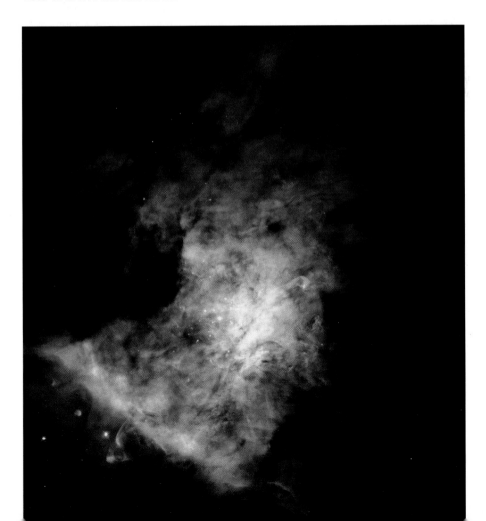

INDEX

Picture Acknowledgements

AAO, photo by David Malin: 149

Alcatel Space: 62

Anglo-Australian Telescope Board: 145, 186

P. Brandt, G. W. Simon and G. Scharmer, Swedish Vacuum Solar Telescope, La Palma, June 1993: 11

Caltech: 175

Cavendish Laboratory: 159

CERN: 181

Peter Convey: 50

2dF Survey: 153, 154

DreamWorks: 124

ESA: 66–7 & 92–3

Galaxy Picture Library: 2 & 28 (Robin Scagell), 6–7 & 19 (Richard Wainscoat), 12 (University of Michigan), 13 (Manchester Astronomical Society), 15, 17 (Robin Scagell), 24–25 & 29 (right, JPL), 28 (Robin Scagell), 29 (left), 30 (left, NASA), 30 (centre, USGS), 31, (NASA), 32–3, 45 (Robin Scagell), 46 (Michael Stecker), 48 (NOAA), 52–3 (JPL), 55 (Malin Space Science Systems), 58 (NASA), 59 (Richard Bizley), 61, 63, 73 (Alain Maury), 74 (Gareth Williams), 75 (both, JPL), 76 (STScI), 78 (John Hopkins University), 79 (John Hopkins University), 81 (Chris Livingstone), 86 (Michael Stecker), 89 (NASA), 90 (NASA), 94–5 & 109 (Don Davis/NASA), 97 (David Brown), 103 (Roger Lynds/AURA/NOAO/NSF), 104 (ESO), 106 (Michael Stecker/Bill Fletcher), 112, (LPI), 132 (NASA), 136, 137 (STScI), 138–9 & 156 (left, NASA), 141 (TRACE/Stanford-Lockheed Institute for Space Research), 144 (Rob McNaught), 152 (Andrew Stewart), 155 (STScI/ESA), 156 (right, Richard Wainscoat), 158 (ESO), 162 (NASA), 163 (NASA/NSF), 168, 170 (STScI), 179 (STScI), 180 (STScI), 182 (NASA/NSF), 184 (STScI).

Mark Garlick: 26, 38, 102, 111

David A Hardy: 34, 42–3 & 56, 118–19 & 129, 135

Hencoup Enterprises: 123, 143, 171

Joint Astronomy Centre, Hawaii: 27

JPL: 37 (Geoffrey Bryden), 125, 126, 140

Malin Space Science Systems: 55

NASA: 146 (R. Sahai & J. Trauger (JPL)), 189 (C.R. O'Dell, Rice University)

Courtesy of the Palermo Observatory: 70

Ian Palmer: 131, 173, 176, 178, 188

Pioneer Productions: 8, 21, 49, 69, 98, 108, 115, 121, 143, 171

ROE/AAT Board: 148

University of Calgary: 82

University of California, Berkeley: 9

Richard Wainscoat/University of Hawaii: 101